W9-BGF-521

a newer world

a newer world

POLITICS, MONEY, TECHNOLOGY, AND WHAT'S
REALLY BEING DONE TO SOLVE THE CLIMATE CRISIS

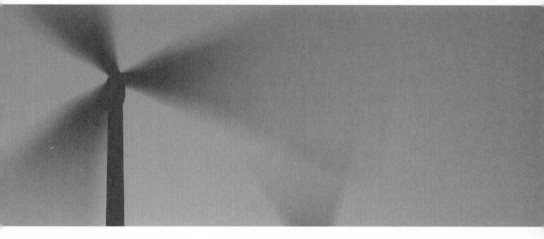

william f. hewitt

FOREWORD BY WILLIAM K. REILLY

University of New Hampshire Press · *Durham, New Hampshire*

University of New Hampshire Press
An imprint of University Press of New England
www.upne.com
© 2013 William F. Hewitt
All rights reserved
Manufactured in the United States of America
Designed by Vicki Kuskowski
Typeset in Granjon by Copperline Book Services, Inc.

University Press of New England is a member of the Green Press
Initiative. The paper used in this book meets their minimum requirement
for recycled paper.

For permission to reproduce any of the material in this book, contact
Permissions, University Press of New England, One Court Street, Suite 250,
Lebanon NH 03766; or visit www.upne.com

Library of Congress Cataloging-in-Publication Data
Hewitt, William F.
A newer world : politics, money, technology, and what's really being done to
solve the climate crisis / William F. Hewitt ; foreword by William K. Reilly.
        p.   cm.
Includes bibliographical references and index.
ISBN 978-1-58465-963-1 (cloth : alk. paper)—ISBN 978-1-61168-351-6 (ebook)
1. Climatic changes—Political aspects.   2. Climatic changes—Economic
aspects.   3. Climate change mitigation.   I. Title.
QC903.H.49 2012
363.738'746—dc23   201215593

5 4 3 2 1

For Bob, and for Gates, two fine gentlemen of the old school.
For Diana, for whom the future should be green.
And for Marian, who not only gets it but has helped me see
so much more than I did before.

Some work of noble note, may yet be done,
Not unbecoming men that strove with Gods.
The lights begin to twinkle from the rocks:
The long day wanes: the slow moon climbs: the deep
Moans round with many voices. Come, my friends,
'Tis not too late to seek a newer world.

—Alfred, Lord Tennyson, "Ulysses"

contents

foreword

I am hardly a disinterested reader of Bill Hewitt's book. I spent several hours with him discussing climate change and the prospects for action. Hewitt has written a seminal work, cogent, compelling, readily accessible to those not steeped in climate lore. He not only explains the rationale for action, but examines options for putting our energy house in order — from accelerated pursuit of efficiency, to protecting forestlands that sequester carbon, to focusing on the role that consumer choices play, and more, much more — so that future generations might enjoy a world as hospitable and productive as we have enjoyed.

Easier said than done, of course. The economic historian John Steele Gordon has observed that a recurring narrative in human history is the search for new, alternative energy sources. With more than $12 trillion invested in current energy infrastructure around the globe, with transportation heavily dependent on petroleum, with demand for electricity likely to grow in the developing world (1.3 billion people lack access to electricity services, and for another billion access is intermittent), we face a formidable challenge to get it right — that is, to reduce emissions of carbon dioxide and other greenhouse gases and to secure in environmentally sound ways the affordable energy essential to support economic growth and meet consumer needs.

As someone with a long career in conservation, I have been aware for some time of the potential for climate change to upend a lot we take for granted. I became immersed in the issue as President George H. W. Bush's nominee to head the Environmental Protection Agency. In his campaign, candidate Bush had promised to bring the White House effect to bear on the greenhouse effect. One of my first briefings, even before confirmation, was on climate change with Frank Press, then-president of the National Academy of Sciences. Soon after came briefings on two EPA reports, one on effects, one on policy measures. The president's secretary of state, James Baker, delivered his first statement as secretary to the relatively new Intergovernmental Panel on Climate Change. He outlined what he called a "no-regrets" policy: measures that made sense in their own right that would also reduce greenhouse gas emissions. From this starting point, we were able to include a phase-out of CFCs in the amendments to the Clean Air Act in 1990. In the last year of President Bush's term, the framework treaty on climate change was finalized in time for him to sign it at the Rio Earth Summit in June 1992 and secure ratification when he returned home. Although many observers consider the treaty insufficient to address the enormity of the challenge, it is nonetheless the strongest statement we have to date regarding U.S. commitments. It creates a moral if not a legal commitment to return America's emissions of greenhouse gases to the level they were in 1990.

Back then, the science of climate change was driven largely by modeling. Today, the science is there for all to see. The overwhelming number of atmospheric scientists and those in related fields consider climate change real and largely unavoidable even if all greenhouse gas emissions stopped today. Thousands of studies across a range of disciplines document changes under way — expansive wildfires, severe drought, record-setting temperatures, forest die-off, the earlier arrival of spring, migration of plant and bird species, extreme weather events, and more. The current president of the National Academy of Sciences, Ralph Cicerone, has stated we know with great confidence that sea levels are rising, average temperatures are rising, and glaciers are melting. His observations are backed by reports from a dozen or so science academies from around the world, as well as successive assessments by the Intergovernmental Panel on Climate Change.

Which is not to say we know all we need to about climate change. The earth is a complex system, with feedback loops and forces at play outside of earth, forces beyond human control. We know the climate has changed before and will again due to natural variation and those outside forces. Notwithstanding the output of far better models, models that scientists tell us do explain *past* climate variation, we cannot yet forecast with accuracy the pace, magnitude, or differential impacts on a localized scale. We constantly are learning more.

Some observers point to positive impacts — a longer growing season, a boost to vegetation from elevated levels of carbon dioxide in the atmosphere.

For agriculture to thrive in a warmer world, however, water and soil moisture would have to be sufficient, new pests and pathogens controlled, and extreme temperatures mitigated. Research suggests that crops will not fare well if temperatures spike beyond some threshold.

The mounting evidence has been sufficient to convince developed and developing countries alike that their national interest is at stake. Many support an international treaty, one with differential commitments pegged to a country's stage of development. It's not just about imposing a burden on these countries, though: opportunities exist to deploy cost-effective low-carbon and renewable energy technologies in Brazil, China, India, Indonesia, Mexico, and elsewhere in the developing world that can help satisfy aspirations for a higher standard of living while addressing climate change.

Though our country has played a constructive role in international meetings of the parties to the climate convention, a national program to address carbon dioxide emissions remains elusive. We are unlikely to see a new international agreement until the United States acts at home.

President Obama campaigned on getting a climate program in place, and indeed his administration has moved on a number of fronts, notably adopting higher vehicle fuel standards and promoting alternative fuel sources. The high-water mark in the climate debate so far may have come when the House of Representatives passed the Waxman-Markey bill in June 2009, which would have created a market-based cap-and-trade regime modeled on the highly successful program to reduce sulfur dioxide emissions from power plants. Climate legislation stalled in the Senate, however, a victim no doubt of the economic downturn and lack of trust in the financial community, which many thought would manipulate the program at the expense of ordinary Americans who would be paying higher energy prices. Nor did a carbon tax — favored by many economists and business leaders, but which runs afoul of anti-tax fervor in Washington, especially among Republicans — gain traction.

As disappointing as this lack of progress in Washington is, a lot is going on across the country, as Hewitt documents, in states and communities and in the private sector. California has ambitious mitigation initiatives under way, including a cap-and-trade program. Several northeastern states have something comparable. We have much to learn from these programs, as well as from the cap-and-trade experience in the European Union. Chicago offers a leading example of what communities are doing to adapt to a warming world. A company on whose board I have served, DuPont, has ambitious energy and water efficiency goals and a strategy of marketing products designed to serve a warmer world. Fuel switching by utilities from coal to natural gas, the new fuel economy standards, new building efficiency standards, and other measures have dragged the business-as-usual trajectory for carbon dioxide emissions down significantly from what the Energy Information Administration first projected in 1990 for the year 2020. As one example, EPA's Energy Star buildings program has certified more than sixteen thousand commercial buildings, which are said to use 35 percent less energy per square foot than the benchmark average.

Much more remains to be done, especially as the economy picks up and produces more emissions. What is needed, in my view, are a price on carbon emissions that can drive innovation and deployment of lower-carbon energy sources; investments in science and technology; incentives for renewables, such as the production tax credit, that can help fledgling energy technologies gain footing; and commonsense, cost-effective product efficiency standards.

And yet, we seem to lack the political consensus to adopt such a suite of measures. The scientific arguments have been made, the rationale for action laid out. But our political leaders hesitate. Some dismiss the science outright. Economic interests with deep pockets have exploited scientific uncertainties, generating considerable confusion among citizens at large who do not follow the climate debate closely. For many Americans, anxieties abound. Climate change seems a distant threat; other priorities demand our attention. No wonder the political establishment has yet to craft a satisfactory response to climate change.

Until Americans grasp directly, emotionally, the implications of climate change for their families, their communities, their livelihoods, their health and that of the natural resources on which all human activities depend, I do not expect to see congressional action. The impetus may yet emerge because of more extreme weather events, more costly disruptions attributed to climate change, more tangible impacts. We have within ourselves the vision and the ability to change course and pave the way for a prosperous, environmentally sound future for our country and for others. That day will come.

—*William K. Reilly*

introduction

TURNING THE CORNER ON THE CLIMATE CRISIS

There is a perception by many informed and concerned observers that we are losing the battle against climate change. We are told that global warming's momentum is unstoppable, that the greenhouse gases that have been accumulating will inexorably poison the oceans and push the climate system irrevocably out of balance. What's worse is that the public doesn't care, the politics are intractable, the smart money is still on the old ways of doing business, and the science is uncertain. What follows in this book is evidence that this is manifestly not the case. We have been making, in fact, astonishing progress on a number of fronts: from clean tech to sustainable development and even on the politics and policy.

There is, however, no sugar coating on the evidence that climate change is real, dangerous, and going to be around in our lives for, unfortunately, generations to come. We have been wreaking havoc on the planet with our deforestation and the degradation of our agricultural lands and marine environments. We have been heedlessly pumping carbon dioxide by the billions of tons annually into our atmosphere for well over a hundred years. The latest figure for the concentration of CO_2 in the atmosphere is 389 parts per million (ppm),* according to the United Nations. This is a 39 percent increase from the value of 280 ppm before the dawn of the Industrial Revolution in the mid-eighteenth century.[1] This does not take into account the other major greenhouse gases, which have also been steadily on the rise.

The impacts from the climate changes that we have been experiencing are virtually everywhere in evidence: sea-level rise, the melting of glaciers and

---

* This indicates the number of molecules of the gas per million molecules of dry air.

polar ice, droughts in many areas and catastrophic storms in other areas, devastation of forests from insect infestations and fires, dramatic changes in ocean chemistry and temperature that are severely distressing marine life, as well as habitat loss and other pressures for many terrestrial species of both flora and fauna. Economies suffer as a consequence, not only in the vulnerable developing world but also in the high-income societies of the developed world.

As we shall see, though, even amid all the often-stark news, there have been hugely positive developments.

· In the fall of 2009, for instance, a "scandal" broke that wasn't a scandal. But the mainstream media, unimaginatively calling it "Climategate," embraced it. The bottom line on climate science, however, is that it is extraordinarily deep, wide, and thorough. The media too, for that matter, has been getting the story right much more often than it has been getting it wrong.

· Similarly, the story of how fossil fuels have been steadily thickening the blanket of our atmosphere, holding in more and more heat, is, in a word, depressing. What is not widely known is that we find ourselves now in the midst of a startling revolution in clean, renewable power production.

· For well over a hundred years, we have relied on a central power plant model that is enormously wasteful. But we are developing decentralized energy options, "smart grids," breakthroughs in green building and energy efficiency, and other clean tech that promise to radically change how we use energy and other resources.

· We witnessed extraordinary events at the UN climate conference in Copenhagen in December of 2009, with world leaders apparently flailing purposelessly in secret meetings while tens of thousands demonstrated in the streets outside. The real story of Copenhagen, though, was that there was excellent progress on effecting programs to mitigate and adapt to climate change. Meanwhile, we have been seeing tangible progress on the environment and energy around the world since before the Earth Summit in Rio in 1992 at the multilateral, bilateral, and national levels. Many of the Obama administration's initiatives, on both the international and domestic stages, have been pushing the edge of the envelope, to use the old test pilots' expression.

· A flexible and proven system for managing greenhouse gases downward came out of the U.S. House of Representatives in June 2009, but died an ignominious death in the Senate thereafter, the victim of special interests and political ideology. International business and finance have nevertheless been moving vigorously forward to minimize risks from climate change, find new solutions to mitigate greenhouse gases, and enthusiastically embrace profitable new opportunities.

· As the rapidly emerging economies of China, India, South Africa, Brazil, and others are surging ahead, they are using more and more energy, much of it from fossil fuels. Their economic growth, dependent to a great extent on hydrocarbons, could overwhelm the climate system with greenhouse gases. At the same time, these roaring economic engines are fast discovering the manifest economic and environmental benefits of renewable energy, energy efficiency, mass transit, and the other tools of modern societies that make life infinitely more livable.

· Deforestation in the Amazon, the Western Pacific, Southeast Asia, Africa, and even in the northern reaches of Canada has been epidemic for decades. In spite of the massive special-interest influence on politics in these places, however, thousands of local, national, and international organizations have successfully mobilized to reduce the degradation of forests and farmland and indeed to create an entire new paradigm for land use that serves people, their communities, and the planet very well.

· The impacts from climate change are truly disturbing. Today. The prospects for the future, even as we have taken up the critical work of building better ways to do business, are worse. Recognizing this, the world has been mobilizing to adapt to the new climatic conditions we face. At the same time, we have been educating ourselves on the problems and solutions, fighting the good fights in the courts, and, perhaps most important, looking at how we live. Lifestyle is a critical factor in the equation of solving the climate crisis.

Some of my students and others have asked me over the last several years if we are addressing the climate crisis in time and with sufficient force and focus to avoid a planetary catastrophe. I tell them I don't know. What I do know, however, is that we are forging new tools for much more sustainable, much cleaner and smarter ways to live. We have been realizing progress in areas like renewable energy that even ten years ago people in the field would have told you was not possible by now. We have a long way to go, but what we are seeing happen is incontrovertible evidence that there is a path to sustainability, that we can, in the words of the environmental prophet Barry Commoner, make peace with the planet.

Another visionary, Alex Steffen, cofounder of Worldchanging, in regard to geoengineering the climate with such bizarre schemes as massive reflectors in space and bus-size carbon-dioxide ingesting machines, said: "Our goal should be to cool the planet in ways that reinforce and restore the resilience of its natural systems."[2] The many ways that we are finding to do precisely that is, in large part, the story of this book.

Our eventual success in this enterprise is uncertain. It is nothing less, in my view, than an evolutionary exercise. Whatever the outcome, though, we should take tremendous heart in the breakthroughs being fostered by the concerned, focused, and highly motivated scientists and writers, engineers and planners, activists and public policy makers, farmers and foresters, teachers and lawyers, financiers and business executives, ordinary citizens and prophets who have been building, brick by brick, a newer world.

We have enormous, virtually unlimited potential in this, but we must take full advantage. As Brutus said,

> There is a tide in the affairs of men,
> Which, taken at the flood, leads on to fortune;
> Omitted, all the voyage of their life
> Is bound in shallows and in miseries.
> On such a full sea are we now afloat,
> And we must take the current when it serves,
> Or lose our ventures.

— William Shakespeare, *Julius Caesar*

a newer world

1

science, media, and the public

THE MESSAGE OF CLIMATE CHANGE

"Climategate"

In the autumn of 2009, as the world was gearing up for a critical summit in Copenhagen on addressing climate change, e-mails and other data were stolen from a server at the Climatic Research Unit (CRU) in the United Kingdom. The CRU is one of the three key centers where data are continuously gathered from thousands of monitoring stations around the world and analyzed. Some of the stolen e-mails were published by climate change "deniers" who attributed nefarious behavior to the work of the climate scientists involved in these communications, namely that they were fudging data and lying about much of their work.* (Police investigations have not yet uncovered who the hackers were.)

The world media seized on this story and started to print accounts that highlighted the claims of the deniers. The story was dubbed, fatuously and wholly unoriginally, "Climategate." The *Wall Street Journal* made much of it, saying the e-mails showed that there were efforts to hide data.[1] The *New York Times* said the climate researchers "seem so focused on winning the public-relations war that they exaggerate their certitude—and ultimately undermine their own cause."[2] Fox News blasted away with both barrels. Fox's Glenn Beck had this to say: "Deleting e-mails, hiding declines, incorrect data, inadequate systems, redefining scientific peer reviews for their own uses! This is what appears to be going on behind the scenes and literally trillions of dollars of policy decisions are being based on what these guys are telling us. If your gut said,

---

* The term for those who challenge the scientific consensus on climate change was originally "skeptics." Scientists, realizing that their discipline embraces skepticism—a healthy examination of the data and how it is analyzed—came to regard those who refuse to accept the data and its carefully sifted analysis as denying the reality of climate change.

'Wait a minute, this global warming thing sounds like a scam.' Well, I think you're seeing it now."[3] The climate blogosphere was in hyperdrive.

In the United Kingdom and the United States, conservative politicians called for investigations. The Fifteenth Session of the Conference of the Parties to the United Nations Framework Convention on Climate Change took place in Copenhagen in December. It did not produce some of the results that many environmentalists and policy makers had hoped for, and some commentators attributed this, in part at least, to the undermining of confidence in climate science by the CRU e-mail controversy and subsequent media storm. In a follow-up meeting to Copenhagen in January 2010, adding insult to injury, China's most senior negotiator on climate change said that we should keep "an open mind on whether global warming was man-made or the result of natural cycles."[4]

Meanwhile, the Intergovernmental Panel on Climate Change (IPCC), the UN body that has been examining climate change since 1989, and which in 2007 shared the Nobel Peace Prize, came under attack as well. In January 2010, criticism was leveled at the IPCC for an inaccuracy in one part of its Fourth Assessment Report, rolled out in 2007. The IPCC response then was to immediately own up to the error and say the standard review process had not been applied properly. "It has, however, recently come to our attention that a paragraph in the 938-page Working Group II contribution to the underlying assessment refers to poorly substantiated estimates of rate of recession and date for the disappearance of Himalayan glaciers."[5]

As with the CRU e-mails, the error set off a storm of media coverage, with extensive comment from skeptics and, in particular, personal attacks on the integrity of the IPCC chairman, Rajendra Pachauri.

After all the blood had been spilled, all the internal and independent investigations in the United Kingdom and United States proved what the climate scientists at the center of the CRU controversy had maintained all along: that their work was thorough and conformed to the rigorous standards of science, that they had not suppressed or "spun" any data or conclusions, and that, above all, the science supporting the deep concern regarding climate change and its impacts was and is rock solid. One independent panel found "no evidence of any deliberate scientific malpractice in any of the work of the Climatic Research Unit."[6]

Similarly, independent reviews of the work of the IPCC showed no significant errors in how the science was interpreted and reported. The Netherlands Environmental Assessment Agency looked at specific chapters on regional impacts. "Overall the summary conclusions are considered well founded, none have been found to contain any significant errors," said the report.[7] The IPCC had itself requested an independent review by the prestigious InterAcademy Council (IAC), an Amsterdam-based organization of the world's science academies. The IAC report did not whitewash the performance of the IPCC, calling for it to "reform its management structure and strengthen its procedures to handle ever larger and increasingly complex climate assessments as well as the more intense public scrutiny coming from a world grappling with how best to respond to climate change." Although not charged with examining the underlying science, the report did say: "The Committee found that the IPCC assessment process has been successful overall."[8] The IPCC embraced the conclusions of the review panel and will incorporate many of them as it moves forward to the Fifth Assessment Report (AR5) to be published between 2013 and 2014.

Science: The Foundation

What these overhyped media events don't properly reveal is the extraordinary scope and depth of the science that has grown up around climate change over the past few decades. The Climatic Research Unit, for instance, was founded in 1972. The CRU has done extensive work since the late 1970s on compiling and analyzing temperature records. In collaboration with the UK's Met Office, the CRU calculates global-average temperature each month. Tens of thousands of temperature readings are taken across the globe from thousands of stations on land and at sea, daily, and fed to the computers. The CRU and the Met Office receive around 1.5 million readings a month.

At the same time, two other entities — both in the United States — are gathering, compiling, and analyzing this massive data flow. The Goddard Institute for Space Studies, which is part of NASA, and the National Climatic Data Center, which is part of the National Oceanic and Atmospheric Administration (NOAA), publish their results regularly. What these two centers and the CRU have consistently found — in their independent analyses — is a similar trend in global temperature: up.[9]

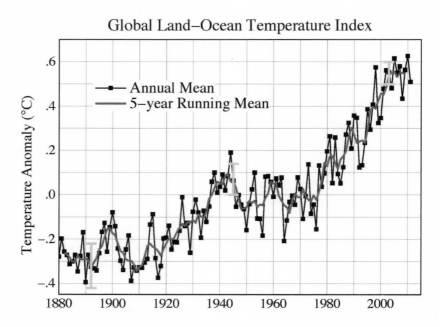

Global Land–Ocean Temperature Index

Temperature Anomaly (°C)

- ■ Annual Mean
- —— 5–year Running Mean

1.1 Global annual mean surface air temperature change. Goddard Institute for Space Studies of the National Aeronautics and Space Administration

Figure 1.1 reflects numbers crunched by NASA. It shows a steady temperature rise, in this case from 1880 to the present. The base period against which the temperature anomaly is measured is 1951–80. The dotted line is the annual mean, and the solid line is the five-year mean.

The three climate science centers work with much more than temperature. They, along with literally hundreds of research institutions and thousands of scientists, are looking at any number of other climate variables, such as precipitation, extent of sea ice and land ice, snow cover, ocean heat content, and moisture levels in soils and forests. All manner of scientific disciplines are involved in studying the often-devastating impacts of climate change: biology, glaciology, atmospheric physics and chemistry, botany, and agronomy, among many others. Climate modelers create sophisticated computer programs that take into account the many and complex variables that go into determining climate, assimilate the data, and make predictions.

When Mount Pinatubo erupted in 1991, for instance, James Hansen, director of NASA's Goddard Institute for Space Studies, and one of the first and still leading climate scientists, accurately predicted the path of the global cooling trend that ensued as a result of the dispersion around the world of the dust from the volcano. He also predicted when the warming trend would resume.[10]

The History of the Science

It was Hansen's testimony before Congress in the hot summer of 1988 that boosted global warming into the public consciousness well beyond where it had been before.[11] "Global Warming Has Begun, Expert Tells Senate" was the lead headline in the *New York Times*.[12] Similar stories appeared around the world.

Michael Oppenheimer, senior scientist for the Environmental Defense Fund in 1988, was sitting immediately to Hansen's left that day. He has followed the science on climate change very closely over the past forty years since his postdoctoral days working on atmospheric physics. For most of that time, he has also been a pivotal member of both the scientific and activist communities working on climate change, stratospheric ozone depletion, acid rain, and other global environmental concerns.

He recalls the progress of the research and the growing interest and concern among scientists. The summer of 1988 was a pivotal time, certainly, but the story had been taking shape for some time before that. In 1979, the National Academy of Sciences produced a report at the request of the Jimmy Carter White House. The "Ad Hoc Study Group on Carbon Dioxide and Climate," chaired by Dr. Jule Charney, a distinguished atmospheric scientist at MIT, reported on the science that had been building from the 1950s and concluded: "For more than a century, we have been aware that changes in the composition of the atmosphere could affect its ability to trap the sun's energy for our benefit. We now have incontrovertible evidence that the atmosphere is indeed changing and that we ourselves contribute to that change. Atmospheric concentrations of carbon dioxide are steadily increasing, and these changes are linked with man's use of fossil fuels and exploitation of the land."[13]

The Charney Report, a landmark to be sure, was however neither the beginning of the focus on global warming and climate change nor the final word. Oppenheimer said: "It's a tricky business to try to identify cause and effect in scientific developments in the broad sense. Something like an academy panel performs a synopsis, an assessment of what happened previously, but also, in some sense, certifies the fact that any new research area is legitimate and interesting."[14]

John Tyndall, the nineteenth-century British physicist, first proved the reality of the greenhouse effect: that certain gases in the atmosphere like water vapor and carbon dioxide trap energy from the sun. Think of the atmosphere as a blanket. The greenhouse effect keeps the blanket thick enough to provide heat enough for life on earth. Anthropogenic greenhouse gases, those resulting from or produced by human beings, like the CO_2 from fossil fuel combustion, make the blanket too thick. Reducing the ability of the climate system to absorb CO_2 through deforestation also contributes to the problem.

In 1896, the Swedish scientist Svante Arrhenius first proposed that carbon dioxide could alter the climate system. In 1938, the lonely voice of British engineer Guy Callendar declared in a paper that CO_2's impact was already in evidence.[15] This is precisely what the Charney panel recognized had become the reality.

The real takeoff in scientific research on climate change, according to Oppenheimer, was in the 1950s, when pioneering scientists like Hans Suess, Roger Revelle, and Charles Keeling picked up the thread and began to look closely at climate change. By the 1960s, scientific research had advanced sufficiently so that in February of 1965 President Lyndon Johnson declared in a message to Congress: "This generation has altered the composition of the atmosphere on a global scale through radioactive materials and a steady increase in carbon dioxide from the burning of fossil fuels."[16]

By the late 1960s and into the 1970s, several early — and by our standards today, quite primitive — computer models of the complex relationships between man-made greenhouse gases and the climate system were being advanced. They were constructed by leading researchers such as Suki Manabe at the Geophysical Fluid Dynamics Laboratory and James Hansen at NASA. Charney and his panel in 1979 drew on their work and that of other pathfinders such as Gordon MacDonald, Bert Bolin, and Veerabhadran Ramanathan.

The Charney panel came into being in the first place because prominent environmentalists Rafe Pomerance and Gus Speth saw the body of scientific evidence building and asked President Carter's science adviser to call on the National Academy of Sciences. Pomerance, Speth, and other activists and scientists, based on the work of the Charney panel, saw the need for another synthesis and persuaded Congress to authorize it, so that in 1983 a comprehensive report was issued by the Carbon Dioxide Assessment Committee of the U.S. National Research Council. This was the most thorough treatment of the subject until the first report of the Intergovernmental Panel on Climate Change in 1990.

As Naomi Oreskes and Erik M. Conway recount in their magisterial 2010 book, *Merchants of Doubt*, the science reflected in the 1983 report was becoming more and more settled — and accepted as such by the climate experts on the National Research Council panel. "The chapters written by the natural scientists were broadly consistent with what other natural scientists had already said. No one challenged the basic claim that warming would occur, with serious physical and biological ramifications."[17]

The final conclusions of the report, however, were different, reflecting the conservative political influence of the panel's chairman, William A. Nierenberg, and economists William Nordhaus and Thomas Schelling. The Reagan White House had a lot of sway as well. The report's synthesis "did not disagree with the scientific facts . . . but it rejected the interpretation of those facts as a problem."[18] One distinguished physicist, Alvin Weinberg, wrote a scathing review of the report's conclusions. Oreskes and Conway describe Weinberg's outrage. He felt that Nierenberg's conclusion "flew in the face of virtually every other scientific analysis of the issue, yet presented almost no evidence to support its radical recommendation to do nothing."[19]

Whatever the political bias of the report's conclusions, it did generate greater interest in the subject and a deeper and broader look at the state of the research. In 1985, a seminal international conference was arranged: The Assessment of the Role of Carbon Dioxide and of Other Greenhouse Gases in Climate Variations and Associated Impacts. Scientists gathered for a week in Villach, Austria, having come from twenty-nine countries under the auspices of the United Nations Environment Program (UNEP), the World Meteorological Organization, and the International Council for Science.

The conference recognized the critical nature of global warming and the role that public policy could have on the progress of warming. It concluded that a number of actions were needed, among them that governments and institutions should undertake considerable further long-term research. It also suggested that a global convention on climate change might be necessary. A comprehensive document summarizing the work of the conference and the major scientific findings to date was compiled by a core group of scientists led by Bert Bolin.[20] Bolin later became the first chairman of the Intergovernmental Panel on Climate Change, serving from 1988 to 1997.

The statement from the Villach conference acknowledged that the problem of climate change was "closely linked with other major environmental issues, such as acid deposition and threats to the Earth's ozone shield, mostly due to changes in the composition of the atmosphere by man's activities."[21] The global nature of many environmental issues was becoming a well-recognized fact. In early 1987, the Report of the World Commission on Environment and Development, titled *Our Common Future*, was published. Gro Harlem Brundtland was the commission's chairwoman. The Brundtland Commission, as it has come to be known, launched the term "sustainable development" into public consciousness, and sustainable development has advanced more and more over the course of the past twenty-five years, entering, in many ways, the mainstream of public policy. "The challenge of finding sustainable development paths ought to provide the impetus—indeed the imperative—for a renewed search for multilateral solutions and a restructured international economic system of co-operation," wrote Brundtland in her preface to the report.[22]

After James Hansen's dramatic testimony in Congress in the summer of 1988, the Toronto Conference on the Changing Atmosphere took place in October and, for the first time, brought scientists together with government representatives to talk about action on climate change. In December, the United Nations General Assembly mandated the formation of the Intergovernmental Panel on Climate Change. Led by Bert Bolin, the IPCC issued its First Assessment Report in 1990.

The IPCC is a collaborative effort by scientists, economists, and other specialists to review research on climate and the impacts of climate change, to analyze it and to synthesize the information, and to make policy recommendations. It does not do original, basic research. Its chairperson and other officers, working with a permanent secretariat, assemble committees of experts to lead the review in specific subject areas. The IPCC has generated four assessment reports so far, with the fifth due to start rolling out in 2013.

In the First Assessment Report, in 1990, the IPCC said: "We are certain of the following: Emissions resulting from human activities are substantially increasing the atmospheric concentrations of the greenhouse gases: carbon dioxide, methane, chlorofluorocarbons (CFCs) and nitrous oxide. These increases will enhance the greenhouse effect, resulting on average in an additional warming of the Earth's surface."[23] In the three assessment reports that have followed, the evidence and the conclusions have been increasingly more dire regarding the state of the climate system.

The IPCC reports are not, however, wholly the province of the scientists and the economists. The final word on what each of the working group reports has to say is determined in negotiations between representatives of national governments. Thus, the IPCC's entire process reflects the Brundtland Commission's call for a "renewed search for multilateral solutions."

International Action

As the body of evidence grew through the 1970s and '80s—and became increasingly irrefutable in its basic conclusions—and as concern grew among environmental nongovernmental organizations and in governments as well, the idea of an international convention to address climate change moved forward. The United Nations Conference on Environment and Development in Rio de Janeiro in 1992—universally known as the Earth Summit—was a gathering of 172 nations, 108 of which were represented by their heads of state or government. President George H. W. Bush of the United States attended. The Earth Summit was characterized by Bush's EPA administrator, William Reilly, as "a watershed event in environmental history."[24] Maurice Strong, the conference secretary-general, called the summit a "historic moment for humanity."[25]

Rio also represented a new level of involvement in global environmental issues by nongovernmental organizations (NGOs). There were nearly twenty-five hundred NGO representatives at the conference itself, and seventeen thousand people attended the parallel NGO Forum. This was by design. Maurice Strong encouraged various interests, including environmental groups, to be involved not only in the summit itself, but in the preparatory meetings as well. Perhaps for the first time on the global scene, the environmental organizations had a major voice and places at the table, seated along with others like business associations and indigenous peoples, as well as the usual suspects from national governments and international governmental organizations. Rio was an exercise in "polycentric decision making."[26]

According to Michael Oppenheimer, the environmental groups were not very focused on global warming until the late 1980s. The groups were only then beginning to come to grips with a problem that had hitherto been seen as not only arcane, particularly for nonscientists, but also a very tough nut to crack on the policy side. But the Montreal Protocol, the international agreement to stop stratospheric ozone depletion, finalized in 1987, proved that the international community could create effective international agreements on the environment. This realization, coupled with what Oppenheimer calls "the spark of public interest" engendered by Hansen's testimony and the other activity on climate change, gave the groups a new sense of mission.[27]

So, as the Earth Summit approached, the U.S. and European groups started to show more interest. Most governments didn't want open access to Rio and similar meetings, but the UN agencies and some governments did want to have the NGOs to help elevate the issue, for the media and for the public. After Rio, there have been annual Conferences of the Parties (COP), and scores of ancillary meetings. South Africa in December 2011 was the site of COP 17. These meetings are important venues for environmental NGOs to meet, network, exchange information, plan strategy and tactics, and brainstorm for the future.

Rio, incidentally, was only the first meeting in the "UN conference decade." It was followed by Vienna in 1993 on human rights, Cairo '94 on population, Copenhagen '95 on social issues, Beijing '95 on women, and Istanbul '96 on habitat. These conferences, the preparatory conferences that preceded them, and the follow-up conferences all brought and continue to bring people together. The people who come are not only the national and international government actors and the big players from the NGO and aid communities, but the grassroots organizations as well.

The Media

As the environmental organizations fully embraced the challenge of confronting what was increasingly being recognized as an ecological crisis of unprecedented magnitude, and the science expanded in depth, breadth, and width, further establishing this perception, publishers and the mass media came to be more interested.

Paul Hawken's *Growing a Business* came out in 1987 and became a seventeen-part PBS series, shown on television in 115 countries. Bill McKibben's *The End of Nature* came out in 1989, Barry Commoner's *Making Peace with the Planet* in 1990, and Al Gore's *Earth in the Balance: Ecology and the Human Spirit* in 1992. All four of these writers have had a long and distinguished track record on investigating and reporting on environmental issues and sustainability. Hawken is an entrepreneur, McKibben is a journalist, Commoner a scientist, and Gore a self-described "recovering politician." All four are activists, and all four have had a wide international following. But, as McKibben himself notes, as influential as these works and others were, while "they opened the conversation, they clearly didn't close the sale."[28]

Not only was 1988 the year of Hansen's dramatic testimony to Congress, the landmark Toronto Conference on the Changing Atmosphere, and the UN's creation of the Intergovernmental Panel on Climate Change, it was also a year of high temperatures and drought throughout North America. The mass media started to become highly aware of climate change, and coverage began to take off. Figure 1.2 shows how newspaper coverage, in the English-speaking world, jumped in 1988 and has subsequently peaked at times of important events relative to the climate change story. As you can see, coverage started to climb steadily in 2003 with the devastating European heat wave that killed scores of thousands of people.[29]

Prominent events raise the profile of an issue. For political scientists this is known as increasing an issue's "salience." High-profile events such as Hansen's testimony, coming as it did in a blisteringly hot summer, the first IPCC assessment report in 1990, the Earth Summit in 1992, the Kyoto conference in 1997, George W. Bush's denunciation in 2001 of the Kyoto Protocol, the European heat wave in 2003, the massive hurricanes Katrina in 2005 and Ike in 2008, the torrential rainstorms and severe flooding in Britain and Ireland in the fall of 2009 all added to the general public's consciousness of—and concern about—climate change. In the United States in 2011, extreme weather ranged from the ongoing severe drought in Texas to Hurricane Irene, which devastated large areas of New England, to some of the worst spring floods

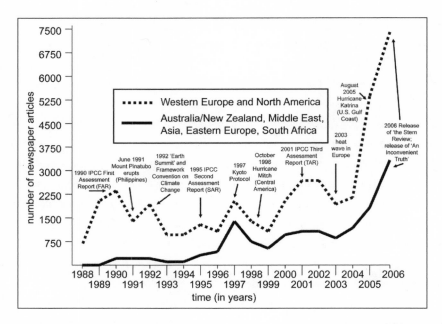

1.2 English-language newspaper coverage of climate change/global warming.
Maxwell T. Boykoff and J. Timmons Roberts, "Media Coverage of Climate Change:
Current Trends, Strengths, Weaknesses," from *Human Development Report 2007/2008*,
United Nations Development Program

in the history of the Midwest. October of 2011 also saw the most severe early
winter snowstorm in New England since before the Civil War.[30] The war in
Iraq, launched in 2003, also heightened concern in the United States and the
United Kingdom about energy security.

The *New Yorker* published a series of articles by Elizabeth Kolbert in the
spring of 2005 on climate change that may arguably have had as significant
an impact as Rachel Carson's series on pesticides did in 1962, subsequently
published as a book, *Silent Spring.* Kolbert covered a tremendous amount of
ground in her reporting and made the issue of climate change starkly clear in
many minds that had perhaps been only nebulously aware of it before. The
book that grew out of the magazine series, *Field Notes from a Catastrophe*,
reached even more people.

Former World Bank chief economist Nicholas Stern spearheaded a review commissioned by the UK government, and published in 2006, that highlighted the dire dangers in climate change to the world's economies. This may

not have caught the imagination of the general public, but it certainly was noticed by economic policy makers in national governments, international governmental organizations, and nongovernmental organizations, and in corporate boardrooms.

Elizabeth Kolbert confirms the view that coverage comes in waves. "Mini-events" like these meetings, reports, and phenomenal weather all draw the media to the story, thus further heightening public awareness. She sees the documentary film *An Inconvenient Truth*, and the exposure it got, as a particularly "big moment."[31]

Al Gore, the driving force and star of that blockbuster documentary, wrote in 2009: "Another encouraging development of the past three years has been the beginning of a major shift throughout the world in humankind's awareness of the climate crisis and its connections to the other major challenges we confront."[32]

Public Opinion, Activism, and the World Wide Web

A poll conducted in fifteen nations, most of them developing economies, commissioned by the World Bank and published in 2009, found that majorities in all but one of the countries (Russia) wanted their governments to do more to combat climate change — and would be willing to pay to do it.[33] (The people in Russia may well have changed their tune after the devastating heat wave and fires during the summer of 2010.) In a survey conducted in 2010 in the United States, more than three-quarters of the respondents were in favor of the government limiting greenhouse gases from business.[34] Another poll out at the same time indicated that large percentages of Americans believed in strong action to combat global warming, including the 65 percent who support the signing by the United States of an international treaty that requires very large, even radical, domestic decreases in greenhouse gases.[35]

Many people around the globe are concerned enough to be vocal and active. For instance, 350.org, a group founded by the veteran journalist Bill McKibben, organized a "global work party" in which there were over seven thousand events in 188 countries on October 10, 2010 (10/10/10).[36] In another global mobilization, "Earth Hour" in March of 2010 recorded 128 countries and territories taking part in turning out the lights to symbolize a recognition of the climate crisis and a willingness to confront it.[37] On October 15, 2009, on "Blog Action Day," over ten thousand websites from more than 150 countries helped promote the climate change message. As McKibben cautions, though, "The Internet has allowed us to organize on a large scale with essentially no money. It's not a substitute for action in the real world — we use it to facilitate those actions, and then to spread the images from them."[38]

McKibben has since expanded the reach of his activist organization, organizing mass demonstrations of opposition to a pipeline project, the Keystone XL, that would have further enabled the environmentally ruinous tar sands oil development in western Canada. McKibben and 350.org realized an enormous victory when, on January 18, 2012, President Obama announced that the pipeline project, as presently envisioned, would not be permitted.

Cyberspace and the blogosphere are very important venues for communicating the news and insights on global climate change. Websites like Grist, Climate Progress, SolveClimate, and DeSmogBlog reach hundreds of thousands of readers around the world. Sites like CleanTechnica and GigaOm appeal to tech-minded people. Environmental Leader, Ecosystem Marketplace, and the different "channels" at GreenBiz.com are geared to the business community interested in clean tech and environmental finance. For people interested in the scientific ins and outs, there are authoritative sites like Climate Central and *RealClimate*, and the important publications *Nature*, *Science*, and *Scientific American* all publish online. The various environmental organizations like Greenpeace, the Environmental Defense Fund, the Natural Resources Defense Council, the Sierra Club, and the Pew Center on Global Climate Change all have vibrant web presences reporting in depth on energy and climate, as do government agencies, from the U.S. Department of Energy, NASA, and NOAA, to the White House, and state, provincial, and national governments throughout the world. Think tanks, research centers, and academic institutions like the Worldwatch Institute, the Union of Concerned Scientists, the Earth Institute, the World Resources Institute, the Rocky Mountain Institute, the Potsdam Institute for Climate Impact Research, and hundreds of others all promote their research, helping to educate and mobilize the public. International governmental organizations like the World Bank, the OECD, the UN Development Program (UNDP), the UN Environment Program (UNEP), the

UN Framework Convention on Climate Change (UNFCCC), and the Intergovernmental Panel on Climate Change are all on the web.

The mainstream media—or MSM, as they're sometimes known these days—all also have a very robust online presence on climate, energy, and the environment. The Associated Press, Reuters, the *Guardian*, the *New York Times*, the *Washington Post*, the BBC, the *Financial Times*, NPR, and TV outlets, among many others, not only continue to devote a lot of ink (and air time) on a regular basis to these subjects, but have blogs, newsfeeds, and an ongoing online profile. One obvious reason more and more people all over the planet are growing concerned and active is the information they receive from the web and the ability they have been afforded to communicate in ways that had hitherto been unavailable to them. (As of the beginning of 2012, there were two and a quarter billion Internet users on earth, out of the seven billion of us. That's a 528 percent increase from 2000.[39])

Disinformation

It is, unfortunately, not incidental to the story of how the media has covered climate change, particularly in the twenty years since the first IPCC Assessment Report and the subsequent Earth Summit, to note that a concerted, focused, and well-funded campaign of disinformation has been waged. The attempt to discredit the science, to instill a sense of doubt about the conclusiveness and the extent of the agreement within the scientific community, is a story well told by Naomi Oreskes and Erik Conway in *Merchants of Doubt*. Oreskes looked at 928—10 percent—of all the papers published on climate change in peer-reviewed science journals over a ten-year period. She chose the 928 papers at random. Not one disputed the view that manmade greenhouse gases (GHGs) were causing a catastrophic environmental crisis.[40]

Greenpeace, for one, has published well-documented reports on the funding for climate change denial by ExxonMobil and Koch Industries, among others. Journalists James Hoggan[41] and Ross Gelbspan[42] have also done considerable spadework in uncovering the campaigns mounted by fossil fuel special interests to discredit climate science. Hoggan writes, for instance, that "it's a story of deceit, of poisoning public judgment—of an anti-democratic attack on our political structures and a strategic undermining of the journalistic watchdogs who keep our social institutions honest."[43] Gelbspan says, "The reason most Americans don't know what is happening to the climate is that the oil and coal industries have spent millions of dollars to persuade them global warming isn't happening."[44] Greenpeace notes that the ongoing "campaigns against climate science continue to receive funding from big oil and energy interests—not just ExxonMobil, but a raft of other companies and foundations whose profits are driven by the products that cause global warming."[45]

A prominent public relations consultant, Frank Luntz, wrote a memo in 2000 that was widely circulated among conservatives seeking to debunk climate science and blunt any public policy progress on the issue. "Voters believe there's no consensus about global warming within the scientific community. Should the public come to believe that the scientific issues are settled, their views about global warming will change accordingly. Therefore, you need to continue to make the lack of scientific certainty a primary issue in the debate."[46]

The problem lies in the fact that even though the misinformation and doubt promulgated by the denialists flies directly in the face of the unequivocal evidence produced by scientists over more than thirty years—and wholly accepted as irrefutable within the scientific community—the media has too often taken the misinformation at its face value. At best, the media has continuously opined that there is a "debate" in scientific circles. At worst, they have broadcast the most outrageous of the claims being touted.

As early as 1994, Michael Oppenheimer saw the danger: "What they've done is try to take scientific understanding and put it on the same level with political opinion. After all, if scientific understanding is the same as political opinion, then everybody's opinion is equally valid. There are no facts. And if there are no facts, there is no extra validity to acting on environmental problems than not acting."[47] (As Senator Daniel Patrick Moynihan said: "Everyone is entitled to his own opinion, but not his own facts.")

We enjoy a vast architecture of science—peer-reviewed journals, conferences, research institutions, and graduate schools, plus government, foundation, and corporate funding to support it all. We have come to rely on science to inform us about dangers to public health and the health of our ecosystems, to provide cures for many of our ills, including fixes for the ills we have brought on ourselves through industrial pollution. Further, we rely on science to inform public policy so that we will better know where to devote the finite resources at our disposal for maximum benefit to ourselves and to posterity. Responsible policy makers have long since become confident in the knowledge that they can depend on expert scientific testimony to help guide them.

Science has been called "the most reliable and self-correcting method ever devised by humans for finding empirical truths about the real world."[48] But what happens if this most reliable knowledge is cast into doubt by a concerted campaign of disinformation? What is the result if environmental problems, and more particularly what many scientists consider the ultimate environmental problem, global climate change, is redefined in some discourse as "non-problematic?" The paper "Defeating Kyoto: The Conservative Movement's Impact on U.S. Climate Change Policy" characterizes the disinformation campaign to discredit the findings of thousands of scientists reflecting decades of work as an attempt to "redefine" their findings. It is a redefinition because this science was—and is—well established, thoroughly tested, and universally accepted.[49]

Regarding the media's culpability, Elizabeth Kolbert's view is that there have been some failures by journalists and news organizations in properly reporting climate change, but that because of its complexities and that it has not been high on the national political agenda, and because the news media report on things that are current and high profile, climate change has not been well and truly covered. "I think that the media has contributed to the general sense of it not being an urgent problem because it's not the lead story of the paper every day."[50] Overall, though, she perceives that the quality and the quantity of media coverage of climate change has been on the upswing.[51]

She has also noted the American public's unwillingness to embrace science. There is a "lack of urgency" in which the public doesn't make demands of the press and the politicians. So it becomes incumbent on political leaders to tell the story and for scientists to develop a much greater willingness and ability to communicate the critical information they have.[52]

Antidotes to Denialism

Combating climate denialism is one of the rationales for websites like Climate Central. A conference of top scientists, journalists, and other parties in 2005 identified the need for an authoritative voice on climate issues. Climate Central, a nonprofit organization, came into being in 2008 to "popularize good information about global warming."[53] It is staffed by a number of scientists.

Another key source for lucid information and commentary on climate science, albeit for those interested in some of the more difficult and esoteric questions, is the blog *RealClimate*. It was founded and is run by working climate scientists, among them NASA climate modeler Gavin Schmidt.

Schmidt agrees with Kolbert that reporters need a "news peg." The media sees the news in terms of a new story, and often one that has immediacy, a sense of conflict or sensationalism. Why, though, hasn't the disinformation been better filtered out by the media? In the particular case of the American media, Schmidt identifies a "journalistic reticence for calling people out." He in fact considers this a "gross ethical violation."[54] Kolbert wrote in *Climate Change: Picturing the Science* — a book co-edited by Schmidt — that "the reporter's habit of giving equal time to the opposing sides on any issue is easily exploited, and of course has been by so-called global warming skeptics."[55]

A 2010 *New York Times* article on the defense of climate science by the academy notes, quite accurately, that "the battle is asymmetric, in the sense that scientists feel compelled to support their findings with careful observation and replicable analysis, while their critics are free to make sweeping statements condemning their work as fraudulent."[56] But a letter to the editor in response says: "If this unfair application of different rules and standards is true, doesn't the fault lie more with the media and the practice of journalism than with the scientists?" The writer goes on to say: "In the apparent interest of balanced reporting, equal voice is too often given to those whose opinions have no demonstrable basis in fact. Journalists owe it to their readers to subject the claims of climate skeptics to the same scrutiny that they apply to mainstream science."[57]

This "asymmetric" journalism, however, has been diminishing over time, and as the volume of stories has grown, so has the quality. Social scientist Matthew Nisbet, director of the Climate Shift project at American University's School of Communication, concluded in a comprehensive report that "with the exception of the editorial pages at *The Wall Street Journal*, in 2009 and 2010 the major national news organizations overwhelmingly reflected the consensus view on the reality and causes of climate change."[58]

Overall, the professionalism of environmental journalism has been steadily improving. The Society of Environmental Journalists was founded in 1990 "to strengthen the quality, reach and viability of journalism across all media to advance public understanding of environmental issues." There are fifteen hundred members from North America and twenty-seven countries beyond. At their annual conferences, climate change has been a persistent theme. Similarly, in June 2010, fifteen hundred people from science, politics, business, and the media came from ninety-five countries to Bonn for the annual Deutsche Welle Global Media Forum to discuss climate change and the media along a wide range of subject areas such as religion's role, the influence of social media and films, and the importance of renewables and effective urban mass transit, among many others.[59]

Public Education

There are many social scientists thinking hard about how to reach the various publics in the world to educate them on the science and the impacts of climate change, as well as to motivate them to address the issue. The Yale Project on Climate Change Communication, for instance, has done a number of studies, including gauging the extent of Americans' knowledge of climate change and their stance. Analysis of ongoing research is published periodically in their report, *Global Warming's Six Americas*. In June of 2010, they identified approximately 13 percent of Americans in the category of "alarmed" about warming, and 28 percent as concerned. Dr. Anthony Leiserowitz describes the alarmed group as those most likely to respond to their perceptions by becoming active on the issue — in political science terms, an "issue public." The size of this group in this case makes it comparable to those who are involved in much more visible movements, such as the immigration reform or pro-choice movements. The difference, Leiserowitz says, is that the people alarmed about climate change are not as well organized as the others.[60]

Some groups that would not normally be concerned about macrocosmic issues like global warming may be amenable to a message that speaks to their more immediate concerns. Hunters and anglers, for example, were the subjects of outreach from the four-million-member National Wildlife Federation. Because the presentations focused on local environmental impacts, something that concerned these outdoorsmen quite a bit, there was much greater penetration of the message. As a consequence, the NWF was able to generate signatures from 670 organizations throughout the United States in 2007 for a letter urging Congress to create new climate legislation.[61]

One of the themes to emerge from the Deutsche Welle conference in June 2010 was the idea that media coverage can and often should be constructive, that it can tell a positive story. Elisabeth Rosenthal, for one, has been in the vanguard on this for a few years. Writing for a largely American audience in the pages of the *International Herald Tribune* and the *New York Times*, she has reported on some important lessons learned in Europe, South America, and elsewhere that can be instructive. Her stories have looked at the stunningly successful Bus Rapid Transit system in Bogotá; waste-to-energy programs in Denmark that are highly complementary to recycling; environmentally safe, ultra-insulated passive houses in Germany that require virtually no heating and cooling; and Portugal's quantum leap toward renewable energy.

She wants to "teach by example." The projects on which she has reported have been, in her words, "more successful than people had hoped, not less, and gained public acceptance more quickly than people had hoped, and it doesn't ruin your life." She perceives a notion in the United States that reducing greenhouse gases requires a diminution in quality of life. What she has seen, though, is that people can change how they do some things without any particular loss in their comfort, affluence, or mobility. This is a message that she hopes more Americans can learn and embrace. She wants people who have concerns regarding climate change and the environment to be practical and see what opportunities exist.[62]

At the *New York Times*, at least, there has been a recognition that readers truly enjoy the kind of positive stories that Rosenthal writes. They are often on the list of the "most e-mailed." And, in very difficult times indeed for newspapers and news organizations, the *Times* has been allowing its environmental coverage to grow and prosper.

Common Sense

The explosion in access to and use of the Internet — and the concomitant birth and exponential growth of the blogosphere — have further enhanced the ability of important stories on climate, energy, and clean tech to reach readers. One of the most influential blogs has been *RealClimate*, led by Gavin Schmidt. It has had over ten million visitors in the slightly more than ten years it has been in business. So even in the face of virulent denial, of the manifest reality of the climate crisis and its anthropogenic origins, Schmidt believes "common sense gets through" in a democracy. He recognizes that people are sometimes threatened in their economic, ethical, religious, or political values by science and what it may be telling us, and this fosters the desire to ignore, doubt, or deny what is being recorded and reported.[63] Nevertheless, the inexorable progress of science reaches the policy makers and informs good decision making. If this is not always the case in Washington, it is often the case, and it is certainly much truer in other national capitals and in the halls of international governmental organizations like the UN and the World Bank, and in the C-suites of Fortune 500 companies where the chief executives do their work.

The message of the climate crisis, the necessity of confronting it, and the abundant opportunity for economic development in setting a course to clean tech are also quite resonant in some key localities. Although the U.S. midterm elections in 2010 revealed a stunning current of climate denialism among the Tea Partiers, which became the standard for Republican candidates throughout the country, California voters rejected that theme. The attempt to roll back California's aggressive attempts to confront climate change with greenhouse gas regulations and by advancing renewable energy and other clean tech failed resoundingly at the polls. Californians from all along the political spectrum recognized the moral and intellectual bankruptcy of denying climate change and cutting off efforts to promote economically and environmentally sane programs. In the most populous U.S. state, with one of the largest economies in the world, the voters also returned three very strong environmentalists to office: Jerry Brown as governor, Gavin Newsom as lieutenant governor, and Barbara Boxer as U.S. senator.

The Grand Old Party seems to be the last outlier among major parties in the world's democracies that refuses to acknowledge the manifest reality of climate change. Conservative party leaders from David Cameron and Nicolas Sarkozy to Angela Merkel all have robust climate and energy programs. Even Russia's president at the time, Dmitry Medvedev, in the wake of the catastrophic drought, heat wave, and fires in his country in the summer of

2010, said: "What's happening with the planet's climate right now needs to be a wake-up call to all of us, meaning all heads of state, all heads of social organizations, in order to take a more energetic approach to countering the global changes to the climate."[64]

The Avatar of Climate Change

No single person in the world personifies the message of climate change more than former U.S. vice president Al Gore. He has been, of course, the target of vilification from the American right for decades owing to his staunch environmentalism. President George H. W. Bush called him "ozone man" in the 1992 election campaign. Bush said of Gore, "This guy is so far out in the environmental extreme we'll be up to our necks in owls and outta work for every American."[65] Gore is consistently a punching bag for climate denialists in the blogosphere, on the campaign trails, and in the editorial pages.

Gore, however, has managed to explain the science, highlight the dangers of global warming, clarify the options and opportunities in doing the world's business more sustainably, and generally elevate the discussion as no other individual in the world has been able to do. *An Inconvenient Truth* appeared in 2006. This film documentary showing Gore's presentations on climate change had a tremendous impact worldwide. A study in 2007 showed that 66 percent of those people in forty-seven countries throughout the world who saw *An Inconvenient Truth* said the film had "changed their mind" about global warming; 89 percent said watching the movie increased their awareness.[66] It received the Oscar for best documentary in 2007 and received scores of other film awards. Gore shared the Nobel Peace Prize in 2007 with the Intergovernmental Panel on Climate Change, further illustrating the importance of the issue.

The chairman of the Norwegian Nobel Committee thanked Gore "for his great courage and unremitting struggle!"[67] Gore's Nobel lecture is quite eloquent. Here is a distillation of the message to which he has devoted most of his life and energy, certainly in the last ten years:

We, the human species, are confronting a planetary emergency—a threat to the survival of our civilization that is gathering ominous and destructive potential even as we gather here. But there is hopeful news as well: we have the ability to solve this crisis and avoid the worst — though not all — of its consequences; if we act boldly, decisively and quickly.[68]

Gore has expanded his role in several ways. The Climate Project, for instance, trains people to deliver the story of climate change in much the same way that Gore's slide show and the film did. Gore himself has taught over three thousand people. Gore's activities go beyond his role as an educator. He has created activist organizations as well. The Alliance for Climate Protection and Repower America both mobilize thousands of people to lobby, write, and otherwise seek to move their communities and their elected officials. He has also carved out a large presence in the investment community, serving as a consultant and board member to a number of firms, and creating his own firm, Generation Investment Management.

His book *Our Choice: A Plan to Solve the Climate Crisis* came out in November 2009. Within a year, there were over four hundred thousand copies in print.[69] An increasingly interested and involved public is indicated by the success of Gore's book and scores of others. These books are either describing the depth of the problem or delving into the many, diverse, economical and technologically accessible ways in which we can advance our societies at the same time that we allow the earth to restore its natural systems. Dozens of films that deal with pressing environmental issues, very much including climate change, are being showcased at festivals and through vehicles such as the Sundance Channel. A recent search on YouTube returned 310,000 results for "climate change."

But the rise in consciousness is not, by any stretch, happening just at the level of how the general public, or even the "issue public," regards climate change. Inventors, engineers, entrepreneurs, financiers, investors, and large corporate decision makers from around the world are raising the ante every day on their involvement in the clean-tech revolution that is happening — and not a moment too soon.

green power

RENEWABLE ENERGY COMES INTO ITS OWN

World on Fire

The world is on fire. In fact, it has been on fire for over 250 years. Until the Industrial Revolution, we used wood and dung to fuel our cooking and heating fires, just as billions of people in the still-developing world do today. It's a story that has been well told: James Watt's steam engine in Scotland facilitated the explosion of heavy industries like steel, ship and railroad building, and coal mining. Western Europe, America, and Japan grew rich and powerful. Colonial powers seized vast tracts of Africa, Asia, and Oceania. And we started polluting our planet in earnest. We have pumped nearly 1.3 trillion tons of carbon dioxide into the atmosphere from our fossil fuel use since 1750 (figure 2.1).[1]

Interestingly, it was not industry that first started to kick the climate into a hyperactive greenhouse effect. It was the cutting down of much of the world's virgin forests to facilitate the "Manifest Destiny" of America's inexorable westward expansion and Europe's urban and agricultural explosion. In fact, it was around 1910 when global fossil fuel emissions of carbon first exceeded those from land-use change, and not until around 1950 that they really started to exceed it greatly.[2] (We will look at the reasons for, the impacts from, and the solutions to contemporary deforestation and forest degradation in Chapter 7.)

With the advent of the post–World War II economic boom, greenhouse gases from power plants, surface transportation, industrial agriculture, and manufacturing kicked into high gear as the prime driver of climate change.

The burning of coal, oil, and natural gas, along with the impact of deforestation, has driven the amount of carbon dioxide in the atmosphere from around 260 to 270 parts per million (ppm) in the preindustrial era[3] to 388 ppm

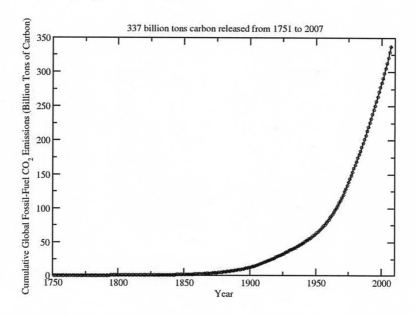

2.1 Cumulative global fossil-fuel CO_2 emissions. Carbon Dioxide Information Analysis Center of the U.S. Department of Energy

in 2010.[4] The burden of CO_2 to the climate system has been steadily increasing, with the largest increases coming after 1945 (figure 2.2).[5]

Coal: A Global Public Enemy

Most of the trouble has come from coal. Through the nineteenth and into the twentieth century, Britain, Germany, France, the United States, and the rest of the industrialized world came increasingly to rely on coal. Vast deposits in these places made coal easily accessible to fire the hearths of industry and domestic use. As Jeff Goodell says in his excellent 2006 book, *Big Coal*, referring to the United States, "It is literally the rock that built America."[6] Then, as the genie of electricity swept through the economies of the industrialized nations, coal was most often the fuel of choice for the central generating facilities. As of the end of the last decade, coal fired 41 percent of the world's electricity. In South Africa, it fueled 93 percent of the electrical power; in China, 79 percent; in India, 69 percent; in Germany, 46 percent.[7] In the United States, coal's use for electrical power has been steadily declin-

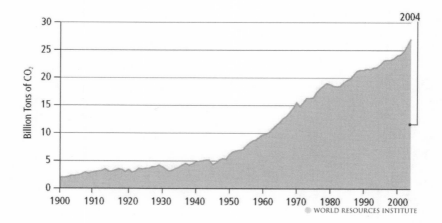

2.2 Global emissions of CO_2 from fossil fuels, 1900–2004. "Navigating the Numbers: Greenhouse Gas Data and International Climate Policy," World Resources Institute

ing in recent years and dipped below 40 percent in late 2011, the lowest share it has had since 1978.[8]

Along with the many remarkable improvements to the quality of life engendered by the Industrial Revolution such as radically improved food production, better freight and human transportation, much-enhanced communication and shelter and heat for hundreds of millions, also came the pollution. Fossil fuels have an amazingly compressed energy potential and an equally astonishing ability to foul the air, water, and land with the waste products of their extraction from the earth, by their passage over land and sea, and particularly in their combustion.

In London, England, during four days in December in 1952, four thousand people died and one hundred thousand were sickened by pollution from coal exacerbated by cold and windless conditions. Thousands more died in the months following as a result of a deadly combination of this one air pollution event and influenza.[9] This event engendered a milestone in the history of air pollution control: the British Clean Air Act in 1956.

A report from Physicians for Social Responsibility enumerates a raft of health impacts at each stage of the coal life cycle — mining, transportation, washing, combustion, and disposing of post-combustion wastes.[10] These include asthma, lung cancer, heart disease, and stroke. One of the most pernicious of the pollutants that comes out of coal-fired power plants is mercury. Mercury is a neurotoxin, and according to the U.S. EPA, coal-fired power plants are responsible for approximately one-third of all mercury emissions attributable to human activity.[11]

In many developing parts of the world where coal is still the primary source of energy, air pollution from coal-fired industrial, electrical, and domestic uses is virtually uncontrolled. China, for instance, gets 79 percent of its electricity from coal, and hundreds of millions of rural Chinese rely on coal and biomass for heating and cooking. One World Bank report concluded that 750,000 people die prematurely in China each year from air pollution.[12]

Coal has other negative environmental and health impacts. Acid rain has devastated forests and freshwater resources in the United States, Canada, and Europe for decades. Thankfully, controls on the sulfur dioxide and nitrogen oxides that spew from coal-fired power plants and are the precursor pollutants to acid rain have greatly reduced the impacts in North America and Europe, but much of that pollution remains unabated elsewhere. Coal mining, of course, is a dirty and dangerous occupation. Among U.S. industries, it is the leading source of death for workers, according to the CDC,[13] and in China, although safety has been increasing steadily, there were still over twenty-five hundred deaths in 2009.[14]

Coal mining and combustion also create massive water pollution. The coal ash that remains after burning is routinely deposited in artificial lagoons. Plants produce about 130 million tons of it every year in the United States alone. On December 22, 2008, a containment dam broke at one such lagoon in Tennessee and released more than a billion gallons of toxic coal ash sludge.[15]

There is an even-greater insult to the lands and waters taking place every day in America's Appalachian Mountains: mountaintop-removal mining. This travesty will have, according to the EPA, destroyed or degraded 11.5 percent of the forests in four states by 2012. Toxic waste will have inundated more than one thousand miles of streams, polluted the drinking water for thousands of people, and caused unprecedented flooding.[16]

Then there is the "resource curse," the phenomenon identified by economists Jeffrey Sachs and Andrew Warner in a landmark paper. "One of the surprising features of modern economic growth is that economies abundant in natural resources have tended to grow slower than economies without substantial natural resources."[17] Jeff Goodell explains it this way: "The development of natural resources tends to crowd out the growth of other, more sustainable industries such as manufacturing."[18]

From being the number one source of greenhouse gas loading to the climate system, to its pervasive health, environmental, and economic impacts, coal is a global public enemy.

Brave New World

Happily, in an astounding number of ways and at an increasingly rapid pace, people are transitioning from what renewable energy pioneer Hermann Scheer called "fuel-driven businesses" to those grounded in technology, those that can build and deploy "energy-conversion" technologies. Another way Scheer described this was in terms of the ownership of the energy. We are "in a shift from commercial primary energies" (coal, oil, gas, uranium) to "noncommercial primary energies" (solar, wind, geothermal, etc.) "and this leads to a broad structural change to the whole economy, and this is a new chance, this is not a burden."[19] What Scheer meant is that there is boundless opportunity inherent in universal access to renewable energy resources by virtue of their being free and limitless, not to mention nonpolluting.

Internationally renowned energy economist Daniel Yergin said in 2008, based on a study by his consultancy, that worldwide clean energy investment could surpass $7 trillion by 2030.[20] Dan Reicher, executive director of the Steyer-Taylor Center for Energy Policy and Finance at Stanford University, agrees that the scale of investment is going to be in the trillions.[21] There are scores of stunning successes in the creation of a brave new energy economy. Wind, both photovoltaic solar power and solar thermal, hydropower, and geothermal are being deployed by the gigawatt (GW).* (According to the U.S. Energy Information Administration, there was 4,012 gigawatts — or a little more than 4 terawatts — of installed electrical generating capacity in the world in 2006.[22]) Marine renewables such as tidal and wave power, hydrogen fuel cells,

---

* A gigawatt is a billion watts — or a thousand megawatts (MW). The watt was named for the Scot James Watt, who brought the steam engine into productive use. A typical incandescent lightbulb uses from 25 to 100 watts. A megawatt is a million watts and is often used to indicate the capacity of a power source.

and other applications for hydrogen, cogeneration, microturbines, even algae as a fuel, are coming on line and becoming more technically and economically viable by the day, advancing, in some cases, at a breakneck pace. These technologies, along with newer, more efficient ways of getting power to consumers, both large and small, and storing it, are critical to the mitigation strategies we are employing to reduce greenhouse gases. They are also important for reducing dangerous air pollution, particularly in the rapidly developing economies of countries like China and India, and in reducing the tremendous environmental and economic consequences of fossil fuel extraction.

The Sky's the Limit: Renewable Energy Potential

Look at this another way, in terms of total global primary energy production, measured in quadrillion British thermal units (Btu).* In 2007, that production was around 475 quadrillion Btu (stated as "475 quad").[23] Now let's convert to another commonly used measure of energy, the exajoule (EJ).[24] One quad equals 1.0551 exajoule. So the 475 quads of total energy produced in the world in 2007 — that is energy derived from oil, coal, gas, nuclear power, hydro, biomass, and other renewables — equals 501 exajoules.

Some researchers presented a paper, "The Potentials of Renewable Energy," at the International Conference for Renewable Energies in Bonn in 2004. In it, they reported that renewables accounted for 62.4 exajoules of power — electricity, heat, and fuels — in 2000. Most of this was generated by hydropower (10 EJ) and biomass (50 EJ).

However, when they looked at the potential using existing technologies, they found over 7,500 EJ a year was available. That is fifteen times the 500 exajoules produced in the world in 2007. Of this potential, 600 was in wind, over 1,600 in solar, and 5,000 in geothermal. (For theoretical potential, they calculated 143 *million* EJ a year!)[25]

We are, in short, able *today* to meet all of our energy needs, for power, heating and cooling, and transportation, using clean, renewable energy. We have but to harness the political will, further enable the technology and build

---

* Quadrillion Btu — 10[15] Btu. A Btu is the quantity of heat required to raise the temperature of one pound of liquid water by one degree Fahrenheit.

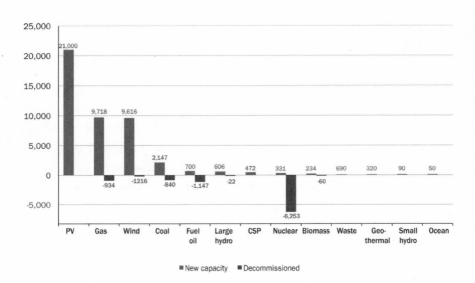

2.3 New installed capacity and decommissioned capacity for 2011 in the EU. European Wind Energy Association, Brussels, 2012

it out, and mobilize the finance. Much of this work is being done, day in and day out, all over the world, and in an increasingly more focused and intensive way.

In Europe, the future has arrived. In 2009, renewables produced 19.9 percent of the EU's electricity consumption.[26] As figure 2.3 illustrates, solar photovoltaic greatly outstripped all other forms of new installed electrical generating capacity, measured in megawatts, in 2011 in the twenty-seven countries of the European Union. Coal and nuclear have been languishing. Natural gas is another winner, as many utilities and industrial concerns have been switching from coal to gas, a fuel that produces half of the carbon dioxide that coal does when it is combusted.[27] But wind and PV have together added more new capacity than gas has each year since 2008.

Globally, nearly 80 GW of renewable capacity was added in 2009, including 31 GW of hydro and 48 GW of non-hydro capacity. China accounted for 37 GW of this, more than any other country. "Green stimulus" packages were a factor in the continuing growth of renewables, but a relatively small percentage of the money committed by governments was spent in 2009, so there is more in the pipeline to help. Two of the East Asian giants, Korea and China, were devoting a big pot to green stimulus.[28] Feed-in tariffs for renewables were also a positive factor, with fifty countries devoting some government money to pay generators of renewable power a premium price.* It is the feed-in tariff that has created such vibrant renewable energy industries in Germany and Spain, among others.[29] In India, like China an explosively developing economy, renewables could reach near 50 gigawatts in the next few years.[30]

A July 2010 study from Greenpeace and the European Renewable Energy Council said that "97% of Europe's electricity and 92% of its total energy use could come from renewables in 2050, cutting CO_2 emissions by 95% with no need for nuclear power or carbon capture and storage."[31] The Germans are echoing the same message, but here it has an official imprimatur: "Electricity supply can be generated completely from renewable energies by 2050 and that secure supply can be guaranteed at all times." This was a statement from Jochen Flasbarth, president of the German Federal Environment Agency.[32] In the UK, they are talking about ZeroCarbonBritain2030, a framework to eliminate all fossil fuel use by 2030.[33]

In the United States, the most famous environmental campaigner in the world, Al Gore, is working, as are a great many others, to create a decarbonized energy economy. Gore's campaign to "Repower America" seeks to have 100 percent of U.S. electricity production come from sources with zero carbon emissions before 2020, a quixotic goal, to be sure, but a good one nevertheless.[34] The U.S. president, Barack Obama, is committed to renewables: "We're accelerating the transition to a clean energy economy and doubling our use of renewable energy sources like wind and solar power — steps that have the potential to create whole new industries and hundreds of thousands of new jobs in America."[35]

Former Japanese prime minister Yasuo Fukuda, at one point not so long ago the leader of the world's second-largest economy, said, "A carbon-free society can no longer be a mere fantasy."[36] It's not.

Wind

One of the constant, limitless, universally available sources of energy that is in the vanguard of the renewable energy revolution is wind power. Onshore

---

* The Intergovernmental Panel on Climate Change defines a feed-in tariff as "the price per unit of electricity that a utility or power supplier has to pay for distributed or renewable electricity fed into the grid by non-utility generators."

and offshore, wind turbines are popping up like mushrooms after the rain, from the United States to Europe to China and most of the rest of the world in between. Wind has, of course, been powering mills for grain and pumping water for crops for many centuries. Wind can also provide electricity by the simple process of turning a turbine, which in turn rotates a generator. This is the way almost all electricity has been generated since Edison and Tesla. However, instead of using steam generated from burning coal or fissioning uranium to turn the turbine, its motion is impelled by rotor blades driven by the wind. The primary energy source is free, does not generate greenhouse gases nor any other pollutants, and will last forever.

Wind power is experiencing exponential growth globally. In 1996, there were 6.1 gigawatts of installed capacity, but by the end of 2010 there were 197 GW. The Global Wind Energy Council expects this to grow to 459 GW by 2015.[37] Offshore is one of the areas where wind is making a huge impact. Turbines have been proliferating along the coasts in Europe, with eighteen new wind farms with 1.6 GW capacity installed in 2009 and 2010.[38] The British alone are developing offshore wind farms that, if all the planned projects move forward, will be able to meet 25 percent of the UK's electricity demand — by 2020.[39] China's offshore wind-power capacity may reach 23 gigawatts by 2020.[40]

In the United States, where offshore has had a very slow start, the Cape Wind project received its final approval from the Department of the Interior in April 2010 — after nine years! Better late than never. This was followed in June by the establishment of the Atlantic Offshore Wind Energy Consortium, a compact between the East Coast states and the U.S. Department of the Interior "to facilitate federal-state cooperation for commercial wind development."[41] Then, in October, Google and a New York investment firm agreed to each take a 37.5 percent stake in a $5 billion transmission spine to connect offshore wind farms along the eastern seaboard.[42] The U.S. secretaries of energy and the interior announced a "National Offshore Wind Strategy" in February 2011, aiming for 10 GW of offshore capacity by 2020 and 54 GW by 2030, from farms along the Pacific, Atlantic, and Gulf coasts, as well as in Hawaiian waters and the Great Lakes.[43]

Here is a sense of the scale: a wind farm with a hundred 1.5 megawatt

(MW) turbines, for example, would have a capacity of 150 MW and could generate enough electricity to power fifty thousand typical American homes.[44]

There are even several well-advanced initiatives to take offshore wind farms even farther out to sea by enabling the turbines to be mounted on floating platforms. The HyWind project, for instance, with a 2.6 MW turbine, piloted by Statoil and Siemens, is floating 10 kilometers off the coast of Norway and is capable of being sited in waters from 120 to 700 meters deep.[45]

There are several advantages to floating the wind turbines instead of anchoring them on platforms. One is obvious: cost. It is less expensive to build a floating platform and to erect the turbine on it at dockside than to do the extensive subsurface testing at sea for pilings to support a fixed platform, build the platform itself, and mount the turbine at sea. There is also considerable time saved — and time, as we know, is money. Also, the farther offshore you go, the stronger the winds are — and the less turbulent.[46]

Yet another plus in siting wind farms farther offshore is that this minimizes or eliminates any concerns regarding visual impacts. The Global Wind Energy Council notes that wind developers are sensitive to these concerns.[47] This has been the most vociferous objection of the vehement opposition to the Cape Wind project in Nantucket Sound. (It is perhaps interesting to note that Bill Koch, the billionaire founder of a fossil fuel energy company, has been a leader of and contributed more than a million dollars to the Alliance to Protect Nantucket Sound, the principal group against Cape Wind.[48]) In any event, it should also be noted, most local residents, near to offshore or onshore facilities, have no objections to them, and, in fact, in some cases wind farms are a tourist attraction.[49] Wind farms are regarded, in most cases, as an economic boon to the communities in which they are sited. The veteran sustainability expert Lester Brown notes a new phenomenon in contrast to the old NIMBY reaction of many communities: PIMBY — Put It in My Backyard.[50]

Analyses of the business potential for wind recognize the extraordinary growth not only in installed capacity but the value added by wind farms all over the world. One prominent research firm predicts that $820 billion will be invested in utility-scale projects over the course of only the next few years, from 2011 through 2017. Beyond that, the industry supports 670,000 jobs worldwide. Wind has arrived in the mainstream for power generation, having achieved "grid parity"* in some markets, a trend that will continue.[51] The Global Wind Energy Council highlights the potential for as many as 2.2 million jobs in the industry by 2020.[52]

The future is bright for wind power, as it is for other renewable energy sources. As Lester Brown put it in a recent book: "The good news is that the shift to renewable energy is occurring at a rate and on a scale that we could not imagine even two years ago."[53]

Here Comes the Sun

According to one advocate, the clean-tech transformation "is in full swing and doesn't need a justification on grounds of climate change. It is simply sound economics to save on scarce costly resources, including energy fuels, to increase energy efficiency and tap resources that are everywhere and abundantly available, such as wind, sun, water and biomass."[54] It's very clear, renewables are here to stay. One of the most dynamic technologies—an array of technologies really—is solar.[†]

There are a number of different solar power technologies, high tech and low tech, deployable on a massive utility-grade scale and on the level of a family in a village in the developing world, and most of these have been growing and are predicted to continue growing, in some cases exponentially. These include photovoltaic (PV), concentrating solar power (CSP), solar thermal, passive solar, and even inexpensive, highly effective solar box cookers. As noted above, researchers have identified 1,600 exajoules of solar energy available for exploitation—using existing technologies. That is three times the amount of energy we use every year worldwide. Another study puts the technical potential for the year 2050 for PV alone at nearly 1,700 EJ and for CSP at 8,000 EJ.[55]

In terms of technology, there are essentially two ways to capture all this sun power. The first is through a photovoltaic cell that receives the sunlight and converts it directly to electricity. These cells use the "photoelectric effect"

---

* Grid parity is generally defined as the threshold at which a renewable energy source supplies electricity to the end user at the same price as conventional grid-supplied power.
† Solar energy, as defined by the Intergovernmental Panel on Climate Change, is "energy from the Sun that is captured either as heat, as light that is converted into chemical energy by natural or artificial photosynthesis, or by photovoltaic panels and converted directly into electricity."

in converting light to energy. Certain materials, such as semiconductors, the most common in use for PV being silicon, are able to absorb light and transfer that energy to electrons. The electrons move along from the cell onto a conducting medium, thus generating electricity. PV has applications on every scale from recharging your cell phone to powering a city. Most of the world's PV goes to small-scale uses, but there is a growing trend toward larger, utility-grade arrays, such as the 500 MW project on the drawing boards now for Edwards Air Force Base in California.

The other main way to harness the sun for electricity is to capture the heat and use it the old-fashioned way: to fire turbines to turn electrical generators. There are a number of technologies available, the main ones being linear concentrator systems, dish/engine systems, and power tower systems. These are generally classed as concentrating solar power (CSP) technologies. They use mirrors to reflect and concentrate sunlight, generating heat that is absorbed by a conducting medium like water, and the thermal energy is then used to create steam to turn a turbine, which turns an electrical generator.

Solar is also, in many cases, not a particularly new or revolutionary technology. We have been using the sun's energy for hot water and heating for over a hundred years.

Obviously, if you are in a good location for sunshine, solar power, whatever the technology, is going to work better and produce more. For utility-scale facilities, though, this is especially important. There are hot spots in the world where we are beginning to see extraordinary developments. Figure 2.4, a map from the World Bank, shows where solar resources are particularly abundant and available.[56] These are the places where concentrating solar power is being developed.

Big facilities are up and running or in the works in a number of places. As of 2010, the global stock of CSP plants was around 1 gigawatt of installed capacity. But this is expected to rise to 25 GW by 2025.[57] One study of an aggressive path to CSP asserts: "With advanced industry development and high levels of energy efficiency, concentrated solar power could meet up to 7% of the world's power needs by 2030 and fully one quarter by 2050."[58]

One of the most exciting prospects in all of this is the Desertec Initiative, a joint venture of European Union countries and companies, with primarily Arab nations from Morocco to Jordan (the so-called MENA area — Middle

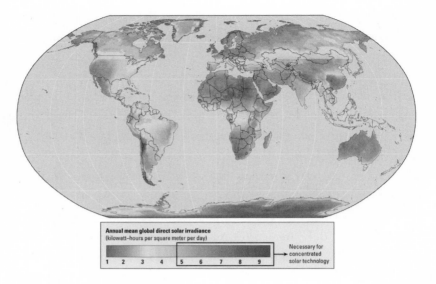

2.4 Global direct normal solar radiation. The World Bank

East and North Africa), and supported by a number of trade associations and international governmental organizations. The goal is to harness the sun and wind in the MENA countries and use that power to supply a large share of the needs of that vast area and also export electricity to Europe to meet as much as 15 percent of its electricity needs by 2050.

This started out as a concept from the Trans-Mediterranean Renewable Energy Cooperation (TREC) in 2003. It has blossomed from a promising idea to a full-fledged industrial initiative with corporate heavyweights like Siemens, Deutsche Bank, Munich RE, ABB, and others behind it, along with many EU and MENA governments fully involved.[59] The European energy commissioner Günther Oettinger has said he expects serious production of electricity for Europe from Desertec by 2015.[60]

One objection to this initiative, particularly from Europeans, many of whom have been nervous about their reliance on natural gas from Russia, might be the risk of relying on electricity imports from states in MENA that have often been at political odds with Europe, even at war, as Egypt was with Britain, France, and Israel in 1956. However, this should be thought of as precisely the kind of relationship between Europe and the MENA states that would create understanding and mutual economic benefit — no less an expert than Rear Admiral Neil Morisetti, climate and energy security envoy of the UK Ministry of Defense and Foreign Commonwealth Office, said just that about Desertec in 2010.[61]

The Desertec concept is being transposed to other venues around the world, including sub-Saharan Africa, India, and Australia. Australia, as you can see from the previous "hot spots" graphic (figure 2.4), has an enormous insolation resource.*

One of the most successful companies to date working on utility-scale solar power is Abengoa, headquartered in Seville. Abengoa has nearly 350 megawatts of solar in commercial operation in Spain and Algeria, with almost 800 MW more under construction, and thousands more MW under development around the world.[62] The cover of this book shows one of their two "power tower" installations at Solucar, west of Seville. Not only does concentrating solar power have enormous potential for providing clean, cheap, and limitless amounts of electricity — it can be stunning to see in operation.

What's not to love? Well, critics of solar power — and of other renewable technologies like wind — make the argument that these sources are "intermittent." That is to say the sun doesn't shine at night and the wind doesn't blow, or at least as much, all the time. Well, that's a reasonable argument as far as it goes — but that's not a very long way. Amory Lovins points out that "the manifest need for some amount of steady, reliable power is met by generating plants *collectively, not individually.* That is, *reliability is a statistical attribute of all the plants on the grid combined.*[63] The consumer relies on the grid and all the capacity that feeds it, not on one facility or the other.

Utilities and energy developers nevertheless recognize the critical need for energy storage and are developing that capacity in tandem with the rollout of utility-scale renewable facilities. Abengoa, for example, among many others, is working on advanced systems to store the heat from their solar plants for dispatch at night. Abengoa, again among others, is also operating and further developing hybrid plants, integrating their solar power production with that provided by coal or natural gas.[64]

Critics of solar power also like to point to the failure of one big company,

---

* Insolation is defined by the U.S. National Snow and Ice Data Center as "the intensity of incoming solar radiation incident on a unit horizontal surface at a specific level."

Solyndra, as an indicator of the weakness of the industry. On August 31, 2011, this major manufacturer of a unique form of solar photovoltaic panels—ones that did not use silicon, as the vast majority of the industry does—filed for bankruptcy. The business had made a big bet on the high cost of silicon for their competitors—and lost.[65] Businesses fail and innovative technologies sometimes don't live up to their promise. The particular problem in this case was that the U.S. Department of Energy had provided Solyndra with a $535 million loan guarantee. With the company's bankruptcy, this is money that the government, unfortunately, is not likely to recover. But does this mean that the solar industry as a whole is destined for failure? On the contrary, one of the main reasons Solyndra failed was that its price was being undercut by the other highly successful manufacturers with which it was competing. Did the company make poor business decisions and the government neglect to perform proper due diligence? Unfortunately, yes. The government has also, at the same time, made billions of dollars in other loan guarantees for companies that have performed quite well, showing a track record that generally exceeds that of the best venture capital firms.

The average installed cost of PV for U.S. consumers has continued to fall: by 17 percent from 2009 to 2010, and another 11 percent through the first six months of 2011. As the Department of Energy reports, "The market for solar PV systems in the United States has grown rapidly over the past decade, as national, state and local governments offered various incentives to expand the solar market and accelerate cost reductions."[66]

One of the biggest projects in the world currently under construction is a three-power-tower complex being built in California's Mojave Desert. It will deliver as much as 392 MW when it is completed in 2013.[67]

The technology, the solid business case, and the dire environmental necessity inherent in the threat of global climate change—all of these are driving solar power toward realizing its extraordinary potential.

Geothermal Energy

Geothermal means heat from the earth. At the earth's core, temperatures exceed those at the surface of the sun. Most of the earth's heat—as much as 90 percent of it—comes from the decaying of radioactive elements contained in the mantle, which sits between the core and the outer crust. The mantle is nearly three thousand kilometers thick (about nineteen hundred miles).

Heat, as we have seen, whether it comes from burning fossil fuels, fissioning uranium, or from the sun, is used to make steam that drives turbines, which rotate electric generators. This is equally true for geothermal power plants. Large geothermal plants, from 1 or 2 MW to the largest complex in the world, the Geysers in northern California, with a capacity of about 725 megawatts of electricity, rely on these very hot temperatures that the earth generates from molten rock, known as magma, below the surface. At the Geysers, the magma is relatively close to the surface, around four miles.

Growth in geothermal electricity production has averaged 5.5 percent over the past thirty years, reaching 10.72 gigawatts at the end of 2010. Because the most easily accessed geothermal resources lie over major tectonic plate boundaries, where earthquakes and volcanoes are concentrated, including along the Pacific's so-called Ring of Fire, it is not surprising that big power plants are sited in California, Mexico, Italy, Iceland, Kenya, Indonesia, and the Philippines.[68] Electricity from geothermal provides a high percentage of the total for several of these countries, including 25 percent for Iceland, with a 17 percent share for Kenya and the Philippines.[69] Japan, also on the Ring of Fire, has enormous opportunity for developing its 23.5 gigawatts of geothermal potential. Another reason to believe that the Japanese are perfectly positioned to take advantage of this resource is the fact that Mitsubishi, Toshiba, and Fuji Electric are among the world's leading suppliers to the geothermal power industry.[70] Japan, in the wake of the nuclear disaster at Fukushima resulting from the March 2011 tsunami — and the rejection of nuclear power by the great majority of its citizens — has a tremendous opportunity. In Iceland, as elsewhere, geothermal is used both for electricity generation and for direct uses. About 90 percent of the space heating in Iceland is from geothermal. The International Geothermal Association expects the installed capacity of power plants to be 160 GW by 2050. Because this is a virtually greenhouse-gas-free technology — with around one-tenth the life-cycle carbon dioxide emissions of coal — and because new Enhanced Geothermal Systems (EGS) are capable of delivering much more power over time, we well may see greater output than the IGA predicts.

EGS uses the powerful drilling tools that have been developed over the years for the oil and gas industry, which allow developers to go down much farther than before and to drill horizontally as well. It also employs the same approach to manipulating the subsurface rock formations that gas developers have been employing to capture hitherto unrecoverable natural gas: fracturing the rock with water pumped down at high pressure. In the case of the geothermal resource, the water is then returned to the surface as superheated steam, which drives the turbine. As with a conventional thermal plant, the steam is then cooled and reinjected into the ground. It is a closed-loop system. The resource is also available virtually everywhere in the world, not just in seismically active zones.

A report from MIT in 2006 estimated that, in the United States, all fifty states have the ability to tap into geothermal using EGS, and that just 2 percent of the heat below the lower forty-eight states could provide 2,500 times the country's annual energy consumption.[71]

Dan Reicher is among those who think that EGS has a tremendous amount of promise. He was the director of climate change and energy initiatives at Google from 2007 to 2010, where EGS is an important focus. It is not an intermittent, or variable, resource, as wind and solar are — it is constant. Using the extraordinary analytical abilities of Google Earth, Reicher saw broadly spread resources across the globe. Google is involved in all four stages of the RDDD pipeline: research, development, demonstration, and deployment.[72]

Like the wind and the sun, the earth's heat is a constant, unlimited, and free resource.

More Renewable Dreams

There are other renewable resources beyond wind, solar, and geothermal that are in the pipeline, some with great promise. Others, like hydro and energy from biomass, are well established and seeing new life in new applications. Natural gas, though a fossil fuel, is also experiencing a renaissance owing to several important factors: its use in power plants creates about half the carbon dioxide that coal does, other pollutant outputs are also reduced — some like sulfur dioxide and particulates radically so — and natural gas is an abundant resource available at an attractive price for the power industry. Natural gas has, along with wind power, been the greatest source of new installed generating capacity in the United States and Europe for several years running.

There have been discoveries in the past several years of huge natural gas deposits in shale formations in the United States, western Canada, Europe, China, and southern Latin America. New technologies will be and are being deployed to extract this gas from the shale.[73] There is, to be sure, considerable controversy, even within the environmental community, regarding the safety of the principal technology being used to extract the gas from shale formations. "Hydraulic fracturing"—also known as hydrofracking or just fracking— uses water and chemicals injected into the rock formations at very high pressure to liberate the trapped gas. Legitimate concerns include proper processing of the wastewater from the fracking operation, threats to aquifers used for drinking water and agriculture, and the release of methane, a potent greenhouse gas, into the atmosphere.

Nevertheless, given the full application of pollution control technology and its mandated use by regulatory authorities to obviate these concerns, many environmental advocates now see natural gas as the key "transition fuel" to a wholly renewable energy economy in the future, an approach that the environmental prophet Barry Commoner called for over thirty years ago in his book *The Politics of Energy*.

Marine Energy

One of the newer and more exciting entrants to the renewable-energy field of play is marine energy. There are a number of ways to harness the power of waves and tides. These include tidal turbines that capture the power of the ocean just as wind turbines capture the power of the wind; barrages that, like dams, hold the water being moved by the tides and then release it for power; buoys that bob on the waves and thus drive generators; and even a "sea snake" that captures the energy of waves as they pass through its cylindrical sections. Other ways to use the ocean to generate power can capitalize on the difference in salinity between seawater and river water or the difference in temperature at different depths. According to the IPCC's *Special Report on Renewable Energy Sources and Climate Change Mitigation*, the theoretical potential for ocean energy technologies is 7,400 exajoules a year, "well exceeding current and future human energy needs."[74] As noted above, we are currently using about 500 exajoules a year. As the IPCC also notes, ocean energy systems, although in their infancy, "may progress rapidly given the number of technology demonstrations," and government policies are helping accelerate implementation.[75] Researchers, entrepreneurs, and investors in government labs and agencies, international governmental organizations and nongovernmental organizations, start-up companies and major industrial concerns, in banks, funds, foundations, and venture capital outfits, are deeply involved in exploring and rolling out marine energy technologies. One comprehensive report from the cleantech industry analysts at Pike Research puts ocean energy at 200 GW by 2025.[76]

In the United Kingdom, many are particularly enthusiastic about the possibilities for both power generation and the creation of expertise and new manufacturing industries for export. The Crown Estate, the entity that manages the extensive property holdings of the British monarchs, owns the UK seabed out to the twelve-nautical-mile territorial limit and over 55 percent of the foreshore. It is the Crown Estate that is leasing the areas offshore for wind power from which, according to the plan, the UK will derive 25 percent of its electricity by 2020. It is also leasing areas off Scotland where 1.2 GW of marine energy will be generated by 2020, half from tidal and half from waves.[77]

For Australia, one recent study put the amount of energy contained in waves on the continent's southern shores at capable of providing three times more than the Australians now use. Several firms are engaged in developing their technologies on these waves.[78]

In Brittany, on the French coast, the Rance River station has been generating an average of 96 MW since 1966. Its top capacity is 240 MW. In 2009, the Koreans brought on line a new station that eclipses the Rance facility, with a 254 MW capacity.[79] Beyond that, the Koreans are on track to build a $3.5 billion, 1.32 GW plant on Incheon Bay immediately adjacent to their third-largest city.[80]

As with wind, solar, and geothermal, the power in our oceans is boundless, carbon free, and there for the taking. This form of power generation, still in its infancy, promises much for the future.

Hydropower

Water also generates electricity in traditional hydropower settings. Freshwater flow is captured behind a dam and harnessed to turn turbines that are connected to generators. Eight of the ten largest power plants in the world are hydroelectric.[81] Worldwide, hydro generates 16 percent of our electricity, nearly 90 percent of the total amount provided by renewables.[82] There were 926 GW of installed capacity around the world as of 2009, with a technical potential for four times that amount. Africa has enormous potential for new hydroelectric capacity, as do Asia and Latin America.

Beyond the ability to build a lot of new conventional capacity based on the technical potential that exists on natural waterways where there is a significant elevation change, only a quarter of the world's existing forty-five thousand large dams have power-generating capacity. Increasing efficiency at existing hydroelectric plants and adding power-generating capacity at dams that don't have any are two big-ticket ideas. In fact, the National Hydropower Association estimates that in the United States alone, 9 GW of added capacity could come from modernization projects and 10 GW from converting non-powered dams. So, there is further unrealized global potential in that we could be upgrading or retrofitting many of these dams with turbines and generators.[83]

Although hydropower is a true renewable resource, there is considerable controversy regarding the environmental impacts of these dams, including the life-cycle greenhouse gas emissions. If, for instance, forests and other vegetation are not cleared prior to the impoundment of the water, then the vegetation will decay and generate large amounts of methane—a greenhouse gas twenty-five times more potent than carbon dioxide. There is, on the other hand, some research to indicate that the overall net GHG impact of hydroelectric reservoirs is greatly reduced because, for one thing, reservoirs are a "sink" for carbon dioxide.[84]

The gargantuan quantities of cement and steel used in the bigger plants is another concern relative to GHG emissions. Hydropower can also cause disruption to the ecology of river systems. The issue of the proper compensation and relocation of people flooded by these facilities is another major problem, spotlighted by a number of projects, most famously by the world's biggest power plant, Three Gorges in China.[85]

Ironically, hydropower plants have become subject to one of the principal regional impacts of climate change: reduced rainfall and glacial mass. Because hydroelectric plants require a relatively high level of water in their reservoirs to function at optimal capacities, when water stress hits a region, the dams have to cut back on their output.[86]

Notwithstanding these concerns, hydropower continues to be a key component of power output, with a number of salient economic and environmental benefits. These include the fact that hydroelectric supports the build-out of some renewables like solar and wind because it can back these up when their output wanes. In any scenario of a fully decarbonized future, a mix of renewables will be necessary. Of course, water is free, and there's no need to extract it from the earth or process it as we do with fossil fuels, in a series of expensive, usually dirty operations. Neither does the water have to be transported. The generators are located at the power source.

Another system for storing energy and then releasing it back to the grid, a critical component of a new, efficient, and renewable energy economy, is pumped storage. Water is pumped uphill during periods of low power demand, then released through turbines when demand rises. Grid storage allows for much greater flexibility in the use of "intermittent" energy sources like solar and wind. Developers in the United States have 31 GW of pumped storage in mind for the immediate future.[87]

Internationally, projects on the grand scale to "run-of-the-river" to microhydro are moving forward, with huge implications for addressing the climate crisis. The International Hydropower Association estimates that hydro offsets 2.1 billion tons of carbon dioxide emissions each year that would perhaps otherwise have emanated from fossil fuel power plants. It further estimates that development of the rest of the technically realizable hydropower potential would offset another 7 billion tons — more than a third of the world's current total output of CO_2.

Biomass

Another renewable resource with great potential for offsetting greenhouse gas emissions — and in some cases even creating carbon-negative power — is biomass. Garbage, for instance, or municipal solid waste (MSW) as it also known, is a form of biomass. It can productively generate large amounts of heat and power in an environmentally sound manner. MSW is but one of many feedstocks. Agricultural, industrial, and forestry wastes; the scrubwood and dung used by billions of people for cooking, heat and light in rural areas of the developing world; and even landfill gas are all biomass feedstocks that are being used today. Of course, wood is the biomass that fueled our homes and cooked our food for thousands of years — and still does in much of the world. Wood, straw, dung, and the like account for about 30 exajoules of energy for cooking, heating, and lighting in the developing world today.[88] This is not without enormous implications, it must be said, for public health and the environment, as we shall see in Chapter 7.

Another 11.3 EJ of bioenergy is used for electricity, heat, combined heat and power (also known as cogeneration), and for transport. The technical potential exists for geometrically expanding this amount to as much as 500 EJ by 2050.[89]

There has been a burgeoning interest for a number of years in growing biomass for use as fuel or for power production. Vast amounts of acreage are planted in corn in the United States and sugar in Brazil for ethanol, and, increasingly, palm oil from Indonesia and Malaysia is being used as a biofuel. The use of these crops for fuel has, however, seriously exacerbated climate change by putting rainforest in the Amazon and peat forests in Indonesia and Malaysia under enormous pressure, leading to deforestation and forest degradation and the attendant loss of these areas as carbon sinks.[90] When land-use changes are taken into consideration, Indonesia and Brazil rank third and fourth in the world as countries contributing to climate change. We will visit this issue more fully in Chapter 7.

The stresses that this biofuel production puts on land normally used for food production have also been identified as having a direct impact on steep increases in food prices. One study noted that using corn for ethanol increases the price of certain staples like milk, eggs, and cereals from 10 to 30 percent in the United States.[91] In the developing world, already malnourished poor populations are much less able to stand up to the pressure of high prices.

These are critically important considerations, obviously, and scientists and policy makers have been taking them seriously. Cellulosic ethanol, made from wood, grasses, or the nonedible parts of plants, is what is referred to as a "second-generation" biofuel. Crops that don't require particularly fertile land as food crops do have great potential as feedstocks. These include switchgrass, miscanthus, and jatropha. Fast-growing trees and agricultural waste such as corn leaves, stalks and cobs, and bagasse, a residue of sugarcane and sorghum processing, also are growing in importance as feedstocks. One of the hurdles in using grasses and other cellulose-rich feedstocks is getting the chemistry right in the processing, but much research is being devoted to unlocking the codes that will make using these resources cost-effective. Mandates for second-generation biofuel use in transportation by the United States and the EU are among the main drivers of this research.

Right now, globally we can generate about 50 GW from biopower, 8 GW of that coming from the United States. There is vast potential beyond this, using the resource base we have seen and with improved technology for cultivation and processing.[92] Biopower has enormous implications for sustainable development in the global South. The huge expense of importing refined petroleum products is an enormous drag on developing economies. Biomass for power and fuel, grown locally, often on marginal lands, can be a huge boost to fragile economies. Jatropha, for instance, is having a growing positive influence in reducing poverty in Mali — and even helping to retard and sometimes reverse desertification.[93]

Using biomass for biochar is another hugely promising bet. Imagine a system that can:

- (potentially) store billions of tons of carbon in soil for centuries;
- dramatically reduce agricultural waste, forest debris, and some municipal solid waste, thus eliminating the production of greenhouse gases that result from their decomposition;
- generate energy to both power itself and a surplus for use in surface transportation or electricity generation; and
- greatly increase the productivity of agricultural soil, thus reducing the need for expensive and polluting fertilizers.

By pyrolyzing or gasifying the feedstock — two types of thermal degradation of the biomass without any air pollution as a byproduct — you produce the char, which is proving, in all sorts of soils all over the world, to have wonderful properties that enhance agricultural or horticultural productivity.

Growing biomass dedicated to biochar production is another way to go. Using waste or specially grown feedstocks makes the process carbon negative — the carbon dioxide that has been biosequestered in the plant material during photosynthesis stays sequestered. In studies of the ancient Amazonian Indian soils known as *terra preta*, the char has been shown to have retained it high carbon content for, in some cases, many hundreds of years.

Ongoing studies by universities, agricultural agencies, and NGOs appear to be proving the great promise of biochar for decentralized energy, agricultural productivity, and the biosequestration of carbon.[94] One recent study put the biosequestration potential of biochar at nearly two billion tons a year of CO_2 equivalent.[95] (Carbon dioxide equivalent — CO_2eq or CO_2e — is a unit of measurement that compares the global-warming contribution of other greenhouse gases with that of carbon dioxide, for standardization. We will look more closely at this in Chapter 4.)

Algae is an organism with virtually limitless potential for powering biomass power plants and as a transportation fuel. The aviation industry, for instance, is deeply involved in R&D on alternative fuels, including algae. Pratt & Whitney, Rolls-Royce, and General Electric were all involved in successful tests in early 2009 in which algae was part of a mix used to fly a big commercial jet. The Defense Advanced Research Projects Agency has been working on algae for military jet fuel, and the U.S. Air Force has a commitment to finding non-petroleum-based fuels. EADS, the European aerospace giant, has demonstrated the use of 100 percent algae-based fuel in a small plane and is working with IGV GmbH to roll out commercial-scale production.

In Arizona, a local utility, APS, is pioneering a system to use algae to fire an electrical power plant and capture the carbon dioxide from the plant to help grow the feedstock. The U.S. Department of Energy has provided some funding with a $70 million grant. This closed-loop system is a form of carbon recycling.[96] The Venetians are one step ahead. They are building a 40 MW power plant for their seaport that will also feed the algae with CO_2.[97] This is real-world, zero-carbon use of algae, and others are following suit, including ExxonMobil and its partner Synthetic Genomics. They are vigorously pursuing well-funded research in the hope of creating a full-scale commercial production capability.[98]

Where Are We Going?

It is abundantly clear that renewable resources can provide all our energy needs: for electricity, transportation, agriculture, for power for industry — heavy and light — and can even provide feedstocks for the thousands of manufactured products we consume all over the world everyday. In doing so, we would avoid the enormous, crushing burden of pollution that so threatens the lives and livelihoods of everyone on the planet today and for the future. If we don't make the transition, we are going to be, to put it politely, in the soup.

Echoing what Hermann Scheer says regarding our shift to "non-commercial primary energies" such as solar, wind, and geothermal, energy *maha guru* Amory Lovins said in the fall of 2009, "The Renewable Revolution has been won. Sorry, if you missed it."[99] As we are seeing, this is not hyperbole. A report from Greenpeace, *The Silent Energy [R]evolution: 20 Years in the Making*, documents the fact that 26 percent of new power plant capacity added to grids around the world in the decade following the new millennium was renewable, with wind power leading the way.[100]

We appear to be in the midst of a massive, global, seemingly inexorable push to clean tech. It will take focus, commitment, and money to stay on track, but the signs are truly auspicious.

bringing it all back home

CLEAN TECH IN ACTION

Electricity, without question, has brought humankind enormous benefits: light and appropriately moderated heating and cooling, refrigeration, cleaning, extraordinary power and flexibility in manufacturing the many products upon which we rely to make life more manageable and pleasant, and, in recent years, the ability to gain access to knowledge and to communicate at speeds and in ways that were undreamed of only a generation ago. Thomas Edison's first power plant, coal fired, was built in downtown Manhattan in 1882 and provided electricity for lightbulbs. Aside from electric power, it also generated, from the start, an enormous burden of air pollution, which made local residents very upset. As we have seen, the constituents of the pollution from fossil-fuel-fired electric power plants not only substantially drive climate change but also seriously degrade the quality of life for many that it doesn't kill outright.

One of the first things that Edison and his contemporaries, the barons of the early electric utility industry, discovered was that the best way for them to deliver their product and at the same time maximize the return on their capital was to build big power plants, often on the outskirts of towns and cities to minimize air pollution's toll on urban residents, and then send the power down transmission lines to their customers. Later, industries that had initially planned to create their own power on-site to drive their manufacturing facilities found that electricity provided by these utilities had become the cheapest option for them. Henry Ford's adoption of the assembly line as the basis for his production could not have occurred without the provision of massive amounts of cheap electricity. Thus the utility grid, consisting of power plants plus the transmission and distribution systems, was born.

There are several models around the world for how the grid is managed.

In much of the world, investor-owned utilities have sway. Nationalized power companies also have an enormous presence. Cooperatives and municipally owned and operated utilities are another means for delivering power. Given the critical importance of electricity in our developed economies and in the rapidly emerging economies of India, China, Brazil, and others, whatever the entity delivering electricity wants from central and state governments it very often gets, including rate increases.

Aside from the pollution and the undue political influence that enormous utilities may have in countries around the world, one of the big problems with the central power plant paradigm is that an enormous portion of the energy that goes into firing the boilers for the steam turbines that drive the generators is lost as heat. How much? Nearly two-thirds in the case of conventional power plants, whether they be fossil fuel or nuclear. Another several percentage points of power are lost in transmission and distribution.

Figure 3.1 shows, in quadrillion Btu of energy, how much energy in the United States flows to electricity generation in total, and how much is measured as "conversion losses." After this, another 7 percent of the energy that remains as electricity is lost in transmission and distribution.[1] This illustrates the situation in the United States, but the percentages for the losses are essentially the same all over the world.

If central, thermal power plants were more efficient, then obviously they would produce far less carbon dioxide and conventional air pollutants. Strangely, though, these power plants have not measurably improved their efficiency since the time, well over a hundred years ago, of Edison's first dynamo, as they were called back then.

If our systems for delivering power were more efficient, the dire health impacts of air and water pollution, for instance, would be greatly diminished. A group of distinguished economists has quantified the "gross external damages" for the United States from an array of critical industries, including power production. For coal-fired electricity, the damage is more than $50 billion a year. That is more than the value added to the economy from the use of coal for electricity.[2]

Another terrible financial burden to the world's economies is the cost of fossil fuel subsidies. These supports increase the purchasing power of consumers in many of the world's poorest regions but at the same time skew the price

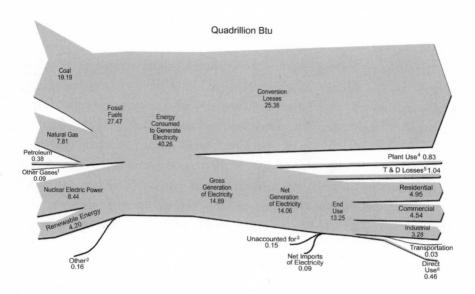

Quadrillion Btu

Coal
19.19

Conversion
Losses
25.38

Fossil
Fuels
27.47

Energy
Consumed
to Generate
Electricity
40.26

Natural Gas
7.81

Petroleum
0.38

Plant Use[4] 0.83

Other Gases[1]
0.09

T & D Losses[5] 1.04

Gross
Generation
of Electricity
14.89

Nuclear Electric Power
8.44

Net
Generation
of Electricity
14.06

End
Use
13.25

Residential
4.95

Renewable Energy
4.20

Commercial
4.54

Industrial
3.28

Unaccounted for[3]
0.15

Other[2]
0.16

Net Imports
of Electricity
0.09

Transportation
0.03

Direct
Use[6]
0.46

3.1 U.S. electricity flow, 2010. Energy Information Administration of the U.S. Department of Energy

of these commodities. The world spends over $550 billion annually on these distorting subsidies, but Fatih Birol, chief economist at the International Energy Agency (IEA), said in 2010 that if we jettisoned subsidies, we would alter the energy equation toward renewables and energy efficiency "quickly and substantially."[3] The IEA further noted more than a year later that "in a period of persistently high energy prices, subsidies represent a significant economic liability." Big oil and gas producers are the biggest culprits: Iran and Saudi Arabia have the biggest subsidies.[4] In the United States, from 2002 through 2008, fossil fuel industries enjoyed $72 billion in supports, while renewables received less than $30 billion.[5] Some of the world's major economic compacts have called for the phasing out of these subsidies. The Organisation for Economic Co-operation and Development (OECD), comprising thirty-four of the world's more advanced economies, has committed to furthering reforms "with the aim of avoiding or removing environmentally harmful policies that might thwart green growth, such as subsidies." The G20, an ad hoc grouping of the world's twenty most influential countries, has said it is necessary to "rationalize and phase out over the medium term inefficient fossil fuel subsidies that encourage wasteful consumption."[6] Easier said than done, perhaps, but the major economies of the world are, after many years, moving in the right direction.

Capital costs for new central power plants and for the creation of the transmission and distribution infrastructure are high. For nuclear power plants, costs are astronomical, which is why no appreciable new nuclear capacity has been added for decades, at least in market economies where private-sector financing has been scarce. In the developed world, infrastructure is aging and, in many cases, has been inadequate to the task of providing enough transmission capacity for new wind farms and other renewable energy resources.

In the developing world, costs are also very high for new infrastructure, and that is one reason why so many people experience "energy poverty." For the hundreds of millions of people who live in rural areas in the world, the great majority of them in sub-Saharan Africa and in South Asia, there simply is no access to electricity. Overall, 1.6 billion people are estimated to live without any access to electric power.[7] Their energy comes from burning biomass — scrub wood and dung, the latter equally if not more valuable as a fertilizer. For hundreds of millions more in the developing nations' cities, the power comes intermittently, often for as little as a couple of hours a day.

The main models for economic development have, for over a century, relied on the provision of centrally generated, relatively cheap electricity, powered for the most part by fossil fuels. The disadvantages of these systems, however, for the climate system, public health, the economies of nations, and indeed continued sustainable momentum toward higher and better standards of living for the world's seven billion people have long since been made clear.

Thankfully, much of the developed world is in a transition to more efficient, cleaner, and more cost-effective modes of delivering power. The developing world at the same time has the unique opportunity to "leapfrog" over the cumbersome and costly old ways of doing business to adopt new ways to power its economies. Renewable energy, generated on a scale that the utility grid can use and feed down the line to its customers, is one way of making this transition. But generating power at the point where it is going to be used is another.

The Decentralization of Energy

"Decentralized energy" (DE), or "distributed generation" (DG) of power, is defined as "electricity production at or near the point of use — irrespective of size, technology or fuel used — both off-grid and on-grid."[8] This can work for homeowners, apartment houses, industrial facilities, institutions like schools and hospitals, office buildings, even military installations. In fact, the U.S. Department of Defense is doing pioneering work in building "microgrids" for their facilities.[9] Why? Simple: Renewable energy can provide a completely self-contained, secure source of power for a base.

DG also overcomes one of the most alarming problems of central thermal power plants (those powered by the heat from burning coal or gas or the fissioning of uranium): massive heat loss. As we have noted, the average energy loss to heat in a plant like this, plus the further loss of power in transmission and distribution, amounts in total to about two-thirds of the energy that has gone into the generating facility in the first place.

It is much more efficient to generate electricity where it is going to be used. Employing natural gas or some other fuel to power turbines right at the location where the power is going to be used enables the facility to also capture the waste heat and use it: for heating, cooling (by using steam to drive the turbine for a chiller unit), and for further powering a secondary power turbine. This practice, known as cogeneration, or combined heat and power (CHP), radically improves the efficiency of energy production and use. State-of-the-art cogeneration plants can reach 80 percent efficiency and beyond.[10] At New York University's cogen plant, the efficiency is 90 percent.[11]

Finland gets 29 percent of its electricity from the waste heat from industrial cogeneration projects and over 75 percent of the heat needed for district heating, with most of the fuel for the plants emanating from the forest products industry.[12]

Beyond this, the Finns and others are even looking to the massive heat from data centers to provide district heating.[13] The powerful and constant activity of the routers, computers, and other high-tech equipment in data processing centers and telecommunications switching facilities generate, as one would suspect, a lot of heat. These facilities require round-the-clock cooling, all year, every year. The air conditioning and the power to run these units are themselves energy-intensive and expensive. How much more efficient then to siphon the heat to productive use heating the buildings in the neighborhood?

In Denmark they take municipal solid waste (MSW)—a fancy term for garbage—and burn it to produce heat and power. Denmark's garbage can generate enough electricity to power 7 million homes. The Danes send 54 percent of their MSW to these facilities, recycle another 42 percent, and only 4 percent goes to landfills.[14]

Denmark is a proving ground for energy efficiency, renewables, and combined heat and power. The island of Samsø, a farming community, produces more electricity, from wind, than it can use (and thus exports the rest to the mainland) and all the heat it needs from district heating plants that burn straw and wood chips. It has become a much-studied laboratory for how to reduce a community's carbon footprint to nearly zero.[15] Samsø is a microcosm, one might argue, of the much, much wider world, yet it is nevertheless emblematic of a transition away from fossil fuels.

Using renewable resources like wind, solar, or geothermal to generate power is real progress. It is, however, even more efficient and cost effective, and it generates more local employment, to decentralize energy production with renewables and high-efficiency systems like cogeneration. In the United States, for example, some renewable energy advocates and local politicians envision a gargantuan complex of wind farms in the Great Plains with transmission lines carrying the power east and west, in some cases for thousands of miles. Long-distance transmission lines, though, are expensive. One expert, Ian Bowles, a former secretary of energy and environmental affairs for Massachusetts, puts it this way: "And of course, the longer the power line, the more expensive it is to build. In New England, we estimate the cost per mile at $2 million to $10 million. The closer electricity is generated to where it's used, the better."[16] These lines are sometimes completely unwelcome in communities along their routes, plus you lose a lot of power, as has been noted, in the transmission and distribution. Because renewable energy resources can be found and utilized everywhere, it is more viable to build power plants locally. Advocates of this approach, like Bowles, also emphasize the importance of local economic development.

Solar at Home

The oil price shocks of the 1970s greatly stimulated the use of solar for the purposes of decentralized energy production. One way to use solar power is to collect and use the heat from the sun's constant bombardment during the day. The International Energy Agency has calculated that the solar thermal collector capacity in operation in fifty-three countries, representing 61 percent of the world's population in 2009, was 172.4 GW, with the jobs created by production, installation, and maintenance estimated to be 270,000 worldwide. The power being produced in this case, it should be noted, is not electricity. It is heat. Only wind power among renewables produces more power at this point.[17]

It's a simple enough process: the sun heats water or another medium such as an antifreeze solution, and then the heated fluid is stored. If it is water that is being stored, then it is used directly for hot water or heating or run through an absorption chiller for cooling purposes. If another fluid is being used to capture and store the heat, then a heat transfer device moves the energy along to the water for the system.

You can use these systems all over the world, in northern climes and southern. The most popular markets are in China, Taiwan, and Japan; Europe; Turkey and Israel; Australia and New Zealand; and the United States. There are generous financial incentives in many places and payback periods on the investment of as little as three or four years.

For the moment, though, photovoltaic is the leading solar technology of choice in Australia. In 2009, for instance, the Aussies installed over 82 MW of PV — more than Spain, a world leader in solar installation — and more than doubled that in 2010.[18] And this PV is largely used in distributed generation so the heat, transmission, and distribution losses from centrally generated power plants are not a consideration.

PV for distributed generation is gaining traction in the United States as well. The U.S. Department of Energy has been helping to support utility-grade renewable energy projects, including solar PV and concentrated solar thermal. For the first time, though, DOE is underwriting the loan for a project that is not a centralized plant. DOE has committed itself to a $1.4 billion loan guarantee for Project Amp, a program that will bring 733 MW of PV to

750 rooftops on commercial buildings in nearly thirty states, with some of the power being used on-site but with the majority being sold to utilities for local distribution.[19] One of the most prominent clean-tech think tanks, the Rocky Mountain Institute, where energy expert Amory Lovins hangs his hat, was over the moon: "Beyond the immediate benefits, Project Amp's sheer size and geographic spread will help scale solar technologies and best practices, further reducing technology and transaction costs for the industry in the future. Additionally, the financial innovation behind Project Amp may help to unlock hundreds of similarly sized projects."[20]

A similar scheme, also backed by DOE, will build as many as 160,000 solar rooftop installations on 124 Department of Defense bases in thirty-three states. The SolarStrong Project will create 371 MW of new capacity. Energy Secretary Steven Chu called it "the largest domestic residential rooftop solar project in history."[21]

How's business overall in the field? Booming, according to the Solar Energy Industries Association. Competition is one of the factors driving growth. The price of solar panels went down by 30 percent from 2010 to 2011. For PV, there was a 69 percent increase in installations in the second quarter of 2011 over the same period in the previous year. The industry was on track to install 1,750 MW of PV in 2011, double that installed in 2010.[22] Does this translate into jobs? You bet. The U.S. solar industry surpassed one hundred thousand workers for the first time during 2011. There was 6.8 percent job growth from August 2010 to August 2011. The industry employs workers in all fifty states.[23]

Worldwide, the numbers are stunning: At the end of 2010, there were 40 GW of installed photovoltaic capacity, 16.6 GW of that coming during that year alone, most of that coming from the European Union. The rest of the world, however, is beginning to fire up its PV. Projections range from 130 to 200 GW of total installed capacity by 2015 — the equivalent of scores of major nuclear power or coal-fired plants.[24] There were 4,012 gigawatts of installed electrical generating capacity in the world in 2006, so 200 GW of that, 5 percent, is considerable. Remember also that much of the installed PV is decentralized, so the 7 percent loss of power in transmission and distribution is obviated.[25]

The future is bright, one might say, for solar power on our roofs and in our backyards.

Aeolus on the Roof: Microwind

Aeolus was the Greek god of the wind. In some parts of Europe, wind power is called aeolic energy. It should not be surprising that another burgeoning technology for decentralized energy is "small wind." Small means 100 kilowatts capacity or less.

There are numerous advantages to the locally generated power of small wind, among them reliability of supply and much lower life-cycle costs for electricity, particularly when you take into account the rising price of power from the utility grid.

Decentralized wind power is certainly nothing new. In fact, we've had windmills grinding grain for flour and powering saws, pumping water from the ground, and helping to control flooding and fostering irrigation, among other uses, for hundreds of years. In Europe, tens of thousands of windmills were in use as late as the early twentieth century. (There were even more water mills in use over time, given the greater reliability of water and its malleability for generating power.)[26] The advent of the Industrial Revolution, though, and steam-powered engines, and later the coming of electrical motors, served to diminish wind and water's role.

But, paralleling the success of its utility-scale big brothers, small wind has been growing too. The new generation of small wind power technologies employs the same general approach as the big turbines but on a much smaller scale. Some wind turbines operate in tandem with other energy sources such as PV or diesel generators as backup.

There are more than 250 manufacturers of small wind capacity, mostly in the United States, Canada, the UK, Germany, China, and Japan, but in twenty-six countries in total. There were 20 MW of new capacity installed in the United States in 2009, a 15 percent increase from the previous year, bringing the total to about 100 MW, still small compared with PV. Manufacturers predict, however, more than 1 GW of cumulative installed capacity in the United States by as early as 2015.[27] Globally, the market for new capacity is growing very fast. One clean-tech industry analyst predicts that the market will increase to $634 million by 2015 from $255 million in 2010. The payback period for these systems is from five to ten years, and they are, in most cases, even more efficient than PV. As small wind gets more exposure and builders and consumers realize the potential, the market will continue to experience excellent growth.[28]

Ground Source Heat Pumps:
Getting the Most from the Earth beneath Us

Another decentralized energy solution that has been changing the shape of the building construction industry is geothermal technology — the ground source heat pump also known as the geothermal heat pump. Like solar thermal systems, ground source heat pumps can provide heating, cooling, and hot water for homes and commercial buildings. And also like solar thermal, the use of heat pumps started to burgeon in the 1970s with the energy crisis.

Unlike their cousins, the utility-scale geothermal plants and the enhanced geothermal systems that are being developed, ground source heat pumps don't require much infrastructure, nor capital costs.

Depending on the location, the ground not far below the surface remains pretty constantly the same temperature: from 45°F (7°C) to 75°F (21°C). In summer, the ground temperature is cooler than the air temperature. Conversely, in winter, the ground is warmer than the ambient air. Ground source heat pumps use this difference in temperature to exchange hot air for cold in the summer and cold air for hot in the winter.

Globally, there are over 130 million pumps in residential buildings and another 15 million in commercial buildings, saving 1.2 percent of global CO_2 emissions annually, according to a report from the International Energy Agency's Heat Pump Program. The potential is even greater: 50 percent of the building sector and 5 percent of the industrial sector's emissions could be saved by heat pumps. That is 1.8 billion tons of CO_2 a year — 8 percent of the total global output.[29]

For every one hundred thousand units of typically sized residential geothermal heat pumps installed, consumers will save about $750 million over the twenty-year life of the equipment. In the United States, over a million pumps are in operation, with tens of thousands of new ones being installed each year. For owners, the energy savings are enormous, with 25 percent to 70 percent lower utility bills, plus these systems increase the value of the property.[30]

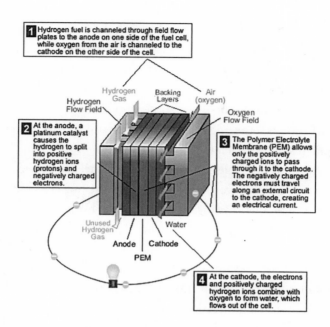

1 Hydrogen fuel is channeled through field flow plates to the anode on one side of the fuel cell, while oxygen from the air is channeled to the cathode on the other side of the cell.

Hydrogen Gas

Backing Layers

Air (oxygen)

Hydrogen Flow Field

Oxygen Flow Field

2 At the anode, a platinum catalyst causes the hydrogen to split into positive hydrogen ions (protons) and negatively charged electrons.

3 The Polymer Electrolyte Membrane (PEM) allows only the positively charged ions to pass through it to the cathode. The negatively charged electrons must travel along an external circuit to the cathode, creating an electrical current.

Unused Hydrogen Gas

Water

Anode | Cathode

PEM

4 At the cathode, the electrons and positively charged hydrogen ions combine with oxygen to form water, which flows out of the cell.

3.2 How fuel cells work. U.S. Department of Energy

How can you miss? Well, if the uptake of this technology is any indication, the answer seems to be "You can't."

Hydrogen and Fuel Cells

A fuel cell uses an energy source to produce electricity, with water and heat as by-products. Pure hydrogen is the best source, but a hydrogen-rich fuel such as methane will do. Applications include primary or backup power for transportation and for remote, off-the-grid locations. Fuel cells also have enormous potential for systems, such as data centers, that require nearly 100 percent reliable power.

In the process, a catalyst separates the hydrogen's charged particles, the negatively charged electrons creating a flow of electricity. The electrons and the positively charged protons then recombine to form hydrogen and, in combination with oxygen, water. Waste heat is the other by-product, which can be used to cogenerate power or can be applied to making ambient heat for a building or hot water. Figure 3.2, from the U.S. Department of Energy, illustrates how it is done.[31]

The first fuel cell was built nearly two hundred years ago. The practical application of fuel cell technology was advanced by the U.S. Apollo program, which eventually landed astronauts on the moon. As long ago as 1970, the year of the first Earth Day, the term "hydrogen economy" was coined to illustrate a vision of clean, cheap, and renewable energy. Certainly, we have not come as far in that time as some might have predicted; still there have been important developments in hydrogen fuel cell deployment.

Hydrogen is the preferred feedstock for a fuel cell because it is more efficient and is pollution-free when it is used in the fuel cell. There are a number of ways to produce hydrogen. Nearly all the hydrogen used in the United States today is extracted from methane. (Methane is the principal constituent, 95 percent, of natural gas.) The process is called "reformation," in which the methane reacts with steam in high temperatures and, in the presence of a metal catalyst, produces the hydrogen with carbon monoxide as a by-product. Another way to make hydrogen is to electrolyze water, separating out the oxygen and hydrogen. In France, nuclear-generated electricity produces the hydrogen via electrolysis.[32] In Iceland, where all the electricity comes from renewable sources—hydroelectric and geothermal—the plan is to produce hydrogen from these renewables, thus enabling a zero-carbon surface transportation fleet running on hydrogen fuel cells.

MIT researchers reported a breakthrough in 2008 in the production and storage of energy combining small renewable energy sources, like solar photovoltaic, with fuel cells in a process that mimics the "essence of plants' energy storage system."[33] In California, one new firm is selling its 100 kilowatt "Bloom Box"—an advanced fuel cell—to Fortune 500 companies like Google, Walmart, and Coca-Cola.[34]

One of the main reasons-to-be for hydrogen fuel cells is that vehicles employing this technology are two to three times more efficient than those using internal combustion engines. These cars have been slow to come to market, but Honda started to roll them out on a test basis in California in 2008. In the fall of 2009, some of the world's biggest automakers agreed to begin selling significant numbers of cars—hundreds of thousands—by 2015, with a concomitant build-out of infrastructure in Europe, with Germany as the starting point, and then on to other parts of the world.[35] A prominent independent research firm has predicted cumulative sales of over a million fuel-cell vehicles

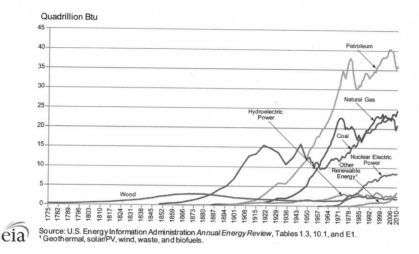

Quadrillion Btu

Source: U.S. Energy Information Administration *Annual Energy Review*, Tables 1.3, 10.1, and E1.
† Geothermal, solar/PV, wind, waste, and biofuels.

3.3 U.S. primary energy consumption estimates by source, 1775–2010. Energy Information Administration of the U.S. Department of Energy

by 2020, generating $17 billion in annual revenue for the carmakers by that time.[36] A public-private consortium of EU and Israeli groups even reported the test flight of a small aircraft using only hydrogen fuel cell power.[37]

The "hydrogen economy" some have dreamed of has not yet come into full flower, but hydrogen and fuel cells are beginning to take an increasingly prominent role in the energy mix of the present and are poised for much more in the near future.

EVs: Plug In, Juice Up, and Go

Fossil fuel use has skyrocketed in the last hundred years, and particularly since the end of World War II. Oil, or petroleum as it is known in most of the world, has been a major component of that increase in fossil fuels — with, of course, an attendant burst in CO_2 loading to the climate system. Figure 3.3 shows how energy has been consumed in America, but is analogous in many respects to what has happened in the other industrial economies as well.

Global figures from 1950 reflect that same steady rise of petroleum consumption, peaking in the early 1970s with the advent of the first of the global oil shocks owing to the Arab embargo, then a plateauing for nearly twenty years, and then a resumption in the rise of consumption.[38]

bringing it all back home

Daniel Yergin's epic history of the oil industry, *The Prize*, recounts how oil came to dominate the world economy in the twentieth century. The discovery of vast reserves of oil, the developing of the technology to efficiently convert oil to useful products — from gasoline to plastic — and the exploitation of this energy-dense fuel by the explosion in use of the internal combustion engine, all served to drive oil's preeminent role in industrialized economies: "One trend in the decades following World War II progressed in a straight and rapidly ascending line — the consumption of oil. If it can be said, in the abstract, that the sun energized the planet, it was oil that now powered its human population, both in its familiar forms as fuel and in the proliferation of new petrochemical products."[39]

With oil's inexorably growing role in powering our surface, sea, and, eventually, air transportation, came the pollution. We have experienced thousands of oil spills, on land and sea, over time, destroying fragile ecosystems and the freshwater and marine resources upon which millions of people depend for their lives and their livelihood. We have despoiled virgin forests in the Ecuadorian Amazon and the Canadian West. We have choked our cities with soot, lead, carbon monoxide, and ozone pollution from the incomplete combustion of petroleum products. We have also spewed many billions of tons of carbon dioxide into the climate system as a consequence of our oil consumption.

Petroleum provides 95 percent of the energy used in the transportation sector. In 2000, transportation was responsible for about 13 percent of greenhouse gas loading globally. In the United States, that percentage is nearly double.[40] We drive a lot of cars in America, and many of them are gas guzzlers. In fact, the internal combustion engine itself is highly inefficient. About 80 percent of the energy that goes into these engines is lost.

But the cavalry is riding to the rescue in the form of not only more efficient hybrid cars, plus the hydrogen fuel cell vehicles noted above, but in the form of vehicles powered wholly, or in large part, by battery-supplied electricity. Vijay V. Vaitheeswaran, a longtime correspondent for the *Economist* and the co-author of ZOOM: *The Global Race to Fuel the Car of the Future*, said there is more innovation in alternative vehicle fuels and supporting technologies going on now than at any time in the past hundred years.[41]

One of the biggest areas of innovation and investment is in electric vehicles. EVs are a very big bet indeed for the future among more and more automakers. One report from the big annual international Detroit auto show in 2010 noted that "the internal-combustion engine seemed almost passé."[42] The Renault-Nissan alliance is predicting sales of between half a million and a million EVs a year in just the next few years—and the companies are well along in building capacity to supply that demand.[43] Tesla, a pioneer in high-end EVs, has partnered with Toyota for a major push into more economical cars. The "goal is to produce increasingly affordable electric cars for mainstream buyers—relentlessly driving down the cost of EVs."[44] General Motors brought its somewhat pricey Volt—MSRP of $41,000—to market in 2011. The car has a range-extending supplementary gas engine and an eight-year/one-hundred-thousand-mile battery warranty.[45] GM is ramping up its production from sixteen thousand in 2011 to sixty thousand in 2012.[46] In China, government subsidies for purchase of EVs and plug-in hybrids, and for recharging infrastructure, lead to the expectation, according to one analysis, that 9 percent of the Chinese passenger fleet will be electric by the end of this decade.[47]

Dan Reicher, the former Google sustainability chief, now at Stanford, is as enthusiastic about this moment as Vijay Vaitheeswaran. Reicher says it's "a very exciting moment for plug-in cars, with bigger opportunities and more practical results now than in the decades we've been talking about EVs."[48] The market research backs up this enthusiasm. EVs generated more than one hundred thousand in cumulative sales through 2011, but will rise to over five million by 2017. The tremendously popular hybrid electric vehicles, like the Toyota Prius, will represent nearly nine million more vehicles sold by then.[49]

Battery range, life, and cost are big concerns, of course, as is recharging. There has been a huge explosion in research and development on batteries in recent years, very much in anticipation of the advent of the electric vehicle era, and the technology has been getting better and cheaper. Government backing, such as "green stimulus" programs, has certainly been a factor. In the United States, $4 billion in public funds are being invested in electric vehicle and battery production. Owing in part to the American Recovery and Reinvestment Act, thirty new battery factories will be in production by 2012, and the United States will have the capacity to produce 40 percent of the world's advanced vehicle batteries by 2015.[50]

So, you may well ask, where do you charge your EV? The immediate obvious answer is at home, preferably at night when electricity rates are cheaper, or at the nearby mass transit hub where you have parked to take a train or bus into work, or in the parking lot at your place of employment. And what if you are on the road? Clearly, we will need new infrastructure for much of this. A number of commercial concerns and public-sector entities are working to bring the necessary technology and infrastructure into play. The roll-out of charging stations parallels the projected growth in electric vehicle sales. Some 7.7 million locations should be up and running by 2017, 1.5 million of those in the United States. Electric vehicle supply equipment prices will drop precipitously over the next few years as both competition and the volume of production increase.[51]

As an example of innovation in this field, Better Place is an excellent choice. It is a visionary company with global ambitions. It has partnerships in Israel, Denmark, Australia, Japan, China, California, Hawaii, and Ontario. It deploys both recharging stations for EVs and a more novel approach: drive-in, drive-out replacement of batteries. The Renault-Nissan Alliance is among the automakers that are working with Better Place to develop cars designed to be able to quickly and easily switch batteries. The company has raised over $550 million to finance its growth, and founder and CEO Shai Agassi makes some breathtaking predictions: one hundred thousand battery-swapping stations in 2013, five hundred million car-charging stations in the United States alone by 2015, and one billion EVs on the road by 2015.[52] Whether or not those numbers come in remains to be seen, of course.

Vaitheeswaran is delighted that Better Place and Renault-Nissan, among the many others involved, are "trying to change both the chicken and the egg" in alternative fuel vehicles, but he maintains that government policy needs to help. In the United States, the federal government has been pushing quite hard, with billions in support for research and development, and tax credits and incentives for the consumer that reach as much as $7,500. Vaitheeswaran cautions, though, that the "power of the incumbent" is the single biggest determinant of whether or not alternative fuels are going to take off—fossil fuel subsidies, in particular, really skew things in favor of carbon-intense energy.[53]

So as bold and hopeful as some of the predictions may be, a lot still depends on further public policy initiatives in the two biggest greenhouse-gas emitting economies, China and the United States, and the rest of the big economies, both developed and developing. Stay tuned.

"Demand-Side Management" and Storage: Managing Clean Energy

Let's cycle back a bit and look at a critical factor in the clean energy equation: how we manage power. This has implications for the production, storage, transmission, distribution, and consumption of energy, and thus for the enormous volume of greenhouse gases for which energy is responsible.

Some of the principal types of renewable energy that we have seen to be growing in use — and exponentially in some cases — are "variable" in their output. Wind is variable depending on the location, time of day and time of year, elevation, and other factors. But wind resources are nevertheless predictable, given the sophisticated modeling tools that are available.[54] The sun, obviously, shines on photovoltaic arrays, CSP facilities, and solar hot-water panels during the day. (Geothermal, traditional hydropower, and marine energy are more constant.)

There is also more demand for electricity during the day as factories and office buildings, schools and hospitals, and many dwellings, not to mention electrified mass transportation systems, are all running full tilt. On sweltering days in an urban environment, air-conditioning demand can be the straw that breaks the camel's back and brings the grid to the brink of collapse.

One way to deal with these variables is to keep demand within acceptable limits. "Demand-side management" is being used by more and more utilities to keep loads within acceptable limits. If a utility is coming to a breaking point, it can, by prior agreement with its customers who have signed on to a program, rely on them to cut back on their energy use. Large commercial users of electricity such as office towers, for instance, will raise the thermostats on their AC to allow the demand to be eased. This process can even be automated, and indeed, in the new age of the smart grid that is dawning, that is precisely how "smart" appliances, including air conditioners, are programmed.

Another way to "smooth" demand is to push back the use of electricity to times when it is more readily available and cheaper as well. One of the potential applications for a broad-based EV infrastructure, for instance, is to have those vehicles use "off-peak" power, generated at night, to recharge. Similarly, appliances such as washing machines can be run at night to take advantage of lower-cost, off-peak prices.

An awareness of how much energy is being consumed is another arrow in the quiver for helping to manage demand. Google PowerMeter, for example, is a free energy monitoring tool that lets consumers view their energy consumption from anywhere, online. With this information, veteran energy activist Dan Reicher has pointed out, people become more conscious of their energy use and consequently make better choices, an example of what has been termed the "Prius Effect." In fact, Google has been working to effect federal policy that will require utilities to provide more real-time access to home energy use, what is being called the "eKnow" bill.[55]

In any event, what utilities have done historically is to deploy a number of different power sources on the grid, feeding in and being used when and where needed. There are base-load plants, and there are peak plants that are run to pick up the slack when demand is high. Peak power production is particularly expensive.

Beyond these methods of smoothing the electric power load, storage is a critical component. There are proven and cost-effective methods of storing power so as to bring it on line when it's needed, hydroelectric pumped storage being the most prevalent. In the United States today, there are 20 GW of this capacity available, with as much as 31 GW on the drawing boards.[56]

Compressed air energy storage is another mode of storing energy produced during off-peak hours for use when demand is high. By using existing storage reservoirs such as salt caverns or abandoned mines, a highly cost-effective approach, you can compress air into the space, then stream it out later to drive turbines, over a good number of hours and at utility scale.[57]

In another innovation, molten salt is being used in one new small CSP power plant in Italy as a medium for storing energy. This has tremendous potential for storing energy for much bigger projects such as Desertec.[58]

The need for storage is also driving the development of utility scale batteries. One 30 MW wind farm in Hawaii is using a large-scale battery energy storage system.[59] Flywheels and ultracapacitors are two other ways of storing massive amounts of energy.

Ice is also a perfectly viable medium for storing energy—and then for use in air conditioning. In fact, over 50 MW of distributed storage using ice is being rolled out in Southern California,[60] and in New York City the Bank of America Tower is using ice for its air conditioning.[61] Bob Fox, one of the principal architects, says for a new build or for a retrofit, "Everybody should be doing ice."[62]

There is even a grid-scale system being piloted for using wind power to synthesize natural gas, which has the advantage of being able to use existing natural gas infrastructure.[63]

There is yet another scheme that would use what may be millions of EVs around the world as batteries. This "vehicle to grid"—V2G—idea relies on the fact that EVs can put out as much as 10 kW, enough power for ten homes.[64]

In any event, Amory Lovins points out that the "latest analyses are suggesting that a well-diversified and well-forecasted mix of variable renewables, integrated with dispatchable renewables and with existing supply- and demand-side grid resources, will probably need less storage or backup than has already been installed to cope with the intermittence of large thermal power stations."[65] As useful and effective as some of these energy storage technologies may be, they can be expensive, and, given the right approach to the design and upgrade of systems, they may be, to a certain extent, superfluous.

Smart Grid

The "smart grid" means different things to different people. It is called by different names, among them the energy internet, the connected electronet, the modern grid, and "IntelliGrid." What is at issue is how best to connect the existing and proposed new generation, storage, transmission, and distribution infrastructure. We have seen how renewables, both at utility scale and decentralized, are burgeoning. We have looked at how demand management and storage can smooth the power grid. We have seen how electric vehicles may well transform surface transportation and electricity use. How do we make all these connections and more, using the cleanest and most advanced technologies, in the most cost-efficient way and in a timely fashion?

Super-grids such as Desertec, the ambitious renewable energy initiative, will require highly efficient long-distance transmission to bring the power generated by the wind and concentrated solar power farms to market not only in the big Middle East and North African urban centers like Cairo, Algiers, Baghdad, and Riyadh, but across the Mediterranean to Europe. Within Europe, a super-grid is already moving forward with the initial move to build an offshore super-grid connecting Norway, Germany, and the UK.[66] Wind farms in West Texas could bring as much as 24 GW of electricity east to the cities, but not without sufficient transmission capacity. In China, the boom is on, with billions being invested in transmission and smart grid technology to manage power more efficiently. Not the least of the investment is for connecting the gigantic Three Gorges project, and other big-ticket renewable developments, to the grid.

Globally, investment in new transmission will reach over $600 billion in the decade from 2010 to 2020. Some of this development will be for upgrading old lines, but much of it will be for new lines to serve the growing demand from consumers and to provide outlets for much of the renewable capacity that is rapidly coming on line. The largest markets for this infrastructure will be the United States, China, and India.[67] Submarine cables will be a big part of the growth, with 60 such high-voltage lines in place in 2011 increasing to 350 by 2020. Europe will account for three-quarters of the total in 2020, owing to its ambitious plans for offshore wind farms.[68]

Another key aspect of the smart grid is the ability to greatly improve the quality of information on grid status, including where and to what extent the system is being stressed. People working on grid operations and planning love the prospect of being able to instantly identify problems. One utility executive noted the smart grid is "just as much about more data and control as it's about new wires." Its value in amassing data will be very important in load forecasting and resource planning.[69] Smart hardware and software working at the transmission and distribution level of the grid will radically cut down the time and expense of power outages, estimated at $100 billion annually in the United States.[70]

Base-load power, as its name suggests, supplies most of the electricity consumed from day to day. What happens, though, when power demand shoots up, such as on a particularly hot day? Auxiliary power is brought on line. But what if you could reduce loads through real-time demand-side management, not to mention increasing the efficiency of consumers and the transmission and distribution of the power? You would need far fewer peak power facilities. As old infrastructure is retired, you will be able to build fewer new assets to replace the old. There is, in fact, already an overcapacity in American power plants because of the inefficiency of the grid. One estimate of the savings inherent in the building out of the smart grid is $400 billion in the United States alone.[71]

Overall, there will be hundreds of billions of dollars spent on smart grid projects over the course of the next several years: for instance, $171 billion from 2011 through 2017 in the Asia-Pacific region[72] and $80 billion from 2010 through 2020 in Europe. Much of that expenditure in Europe will come from the 240 million smart meters that will come on line.[73] Smart meters, employed at the consumer's end, are a key enabling technology for the smart grid, allowing grid operators and electricity consumers to have two-way communication and helping to foster the uptake of decentralized energy by the grid.

Micropower: Energy Self-Sufficiency

One of the most revolutionary aspects of the smart grid is the potential to greatly accelerate the use of "micropower." This is the ability to generate power on a relatively small scale by the home or business owner, from rooftop PV arrays, for instance, or microwind. Largely self-supplying microgrids, where the power is generated within a community or military base or manufacturing facility, can also sell excess power back to the utility grid.

The U.S. Department of Defense has made a big commitment to renewable energy, alternative fuels for their fleets, and microgrids for their bases. In 2011, DOD deployed 38 MW of renewable capacity at its bases. Under a conservative program of building out this capacity, that number will likely grow to 316 MW by 2017, but could go as high as 817 MW by that time if a truly aggressive course is pursued.[74]

The smart grid provides enhanced system reliability and the ability to foster less demand through various means, but it also makes the uptake of distributed renewable energy resources easier. "Net metering" is integral to these efforts because it allows customers with smart meters to sell the power they have generated on-site back to the utility. If and when they need power, the same customers take it from the grid. What has been a one-way flow since Thomas Edison first sent electricity down the line from his coal-fired plant in lower Manhattan in 1882 can now be a two-way flow. As Al Gore says, this "will clearly accelerate the expansion of small-scale generating capacity throughout the grid."[75]

Beyond that, Gore also notes that "the ability on the part of customers to own and operate their own devices for both generating and storing electricity is now clearly growing at a rate that will soon undermine the monopoly model." This is the monopoly that many utilities have enjoyed for over a hundred years.[76] In order to bring the extraordinary energy potential of distributed renewables into full flower, we will have to "decouple" the link between a utility's income and how much power it sells. Regulators are seeing the benefits of restructuring rates to allow utilities to also make money from empowering consumers with energy efficiency and conservation opportunities—what Amory Lovins termed "negawatts"—as well as realizing profits in optimizing the grid and in taking up renewable energy from microgenerators. In California, consumers have realized two dollars in savings for every one dollar invested by utilities in efficiency programs.[77]

Vijay Vaitheeswaran, the energy expert from the *Economist*, calls the smart grid an "enabling technology" because it brings together many aspects of clean technology. He thinks, therefore, that government should have an important role and needs to intervene on regulation and technology. Ideally, we should have a coordinated international approach to facilitate the technology.[78] In the United States, the American Recovery and Reinvestment Act has dedicated more than $4 billion to modernizing the grid.[79] Research, development, and deployment continue to expand here with the U.S. Department of Energy's advanced research labs, the power industry, very much including the industry's Electric Power Research Institute, and major companies like IBM, Google, Accenture, and Siemens, among many others, all working assiduously. In Europe, East Asia, and elsewhere, the smart grid is also fast becoming reality.

Jesse Berst, the executive editor of *SmartGridNews.com*, predicts that the winners of the twenty-first century will be those with reliable and green energy. The "cornerstone of the infrastructure will be the smart grid." He calls the whole framework of renewables, energy efficiency, storage, demand response, microgrids, green building, and even surface transportation tied to the grid the "global smart infrastructure."[80]

Green Building

Another critical component in the clean-tech revolution is the explosion of greener buildings around the world. There are new buildings with all kinds of exciting architectural, design, and engineering features popping up, as well as the perhaps less sexy but no less important trend toward rehabbing and retrofitting existing infrastructure. From green roofs and white roofs; to microgeneration of heat, cooling, and power and stunning efficiency gains; to passive solar design and zero-carbon, zero-waste communities — buildings and the communities in which they are situated are getting cleaner and, in many cases, less expensive to operate.

What was until relatively recently a phenomenon driven by a few visionary architects and embraced by some environmentally minded organizations has become a thriving international movement with professionals and developers, governments and planners, industrial and financial concerns all staking their claims.

The World Business Council on Sustainable Development, for instance, is a CEO-led, global association of some two hundred companies with members drawn from more than thirty-five countries and twenty major industrial sectors; its Energy Efficiency in Buildings project aims to create conditions by 2050 that will allow for all new buildings in the world to "consume zero net energy from external power supplies and produce zero net carbon dioxide emissions while being economically viable to construct and operate."[81]

The Masdar initiative in Abu Dhabi is a sweeping program not only to build a small sustainable high-tech community but also to create an international research network on renewable energy, including a significant graduate school component and a multimillion-dollar clean-tech fund. Masdar — meaning "the source" in Arabic — will be a walled campus of six square kilometers (2.3 square miles) with no cars, no waste, and virtually no carbon output.[82] The first buildings were completed in 2010.

Buildings are responsible for more than 40 percent of global energy use.

They are also responsible for one-third of global greenhouse gas emissions. Buildings also have the most immediate potential for delivering huge energy savings and GHG reductions.[83] McKinsey & Company published a widely used analysis in 2009, *Pathways to a Low-Carbon Economy*. The report asserts that we have the ability to reduce our GHG output by 5 billion tons a year by 2030 from the buildings sector: 3.5 billion in technical potential from efficiency packages plus lighting and lighting controls in new buildings, and 1.5 billion from lifestyle changes like moderating the indoor temperature in cold and hot weather, and conservation measures like using less hot water and turning off appliances and lights.[84] The big news on this is that the reduction in energy use and greenhouse gas emissions from the building sector are and will be highly cost-effective. New York City's Empire State Building, in a major retrofit project, for example, is aiming to reduce energy use by up to 38 percent — and energy costs by $4.4 million annually.[85]

One of the big-ticket items in the Empire State Building retrofit is the windows. There are continual advances being made in glass and window technology, including embedding PV in windowpanes. Other ways to get at energy and greenhouse gas reductions are through insulation, roofing design, heating and cooling technology, water conservation, and on-site power generation, as well as lighting design and technology upgrades including "daylighting" (which is the optimization of natural light within a space). Building-management software is also a key component.

An increasingly prevalent — and profitable — business model is the ESCO, or energy service company. This is a business that develops projects to improve the energy efficiency and maintenance costs for buildings over a period of years, assuming the financial risk that the project will not save the amount of energy guaranteed. This, it turns out, is a rare occurrence.

One of the principal agents of change in the building sector has been the U.S. Green Building Council (USGBC). Founded in 1998, the council is responsible for the influential Leadership in Energy and Environmental Design (LEED) program, which judges just how "green" a building is in its design and construction. There are now 130,000 trained, accredited LEED professionals and nearly five thousand certified projects with another twenty thousand in the process of pursuing certification.[86] LEED is the standard in over a hundred countries, although the British and the French have also promulgated popular and widely used green-building rating systems of their own. Greenbuild, an annual international conference and expo, drew nearly thirty thousand people in 2010.

One of the many benefits of green building that the USGBC and its sister organizations around the world can document is the value for real estate developers. Operating costs routinely decrease and building values increase. Occupancy ratio increases as well, along with the returns on investment.

Bob Fox, one of the principal architects of the Bank of America Tower, the second-tallest building in New York City and one of the greenest buildings in the world, noted its many design innovations, including the use of a gigantic ice machine that provides cooling in hot Manhattan summers. The building uses half the energy of a comparably sized office tower and half the water. The original calculation for the payback period on the water conservation technologies deployed here was twelve years. It turns out to have been fewer than three years.

Because of the many felicitous environmental aspects of the tower, particularly the excellent lighting and ventilation, the occupants truly love working in this building. Fox quotes the building's developer, Douglas Durst, who moved his offices there: "Now I know what it's like to work in a green building. It *feels* good." Al Gore's Generation Investment Management is headquartered there.

The air quality is better than in most hospitals. As one might imagine, this is all quite useful in attracting renters and, for the principal tenant, helps enormously with employee satisfaction and productivity. With cleaner air and other factors in play, such as enhanced control over lighting and temperature, Fox said he calculated that Bank of America will realize $10 million a year as a consequence of increased productivity and employee retention over what it would have had at its former location.[87]

White surfaces reflect the sun's rays, keeping things cooler. It's called the albedo effect.* Houses in hot climates have been whitewashed for centuries.

---

* As defined by the Intergovernmental Panel on Climate Change, albedo is the fraction of solar radiation reflected by a surface or object, often expressed as a percentage. Snow-covered surfaces have a high albedo, the surface albedo of soils ranges from high to low, and vegetation-covered surfaces and oceans have a low albedo. Earth's planetary albedo varies mainly through varying cloudiness, snow, ice, leaf area, and land-cover changes.

This low-tech thinking has come to modern building. In fact, Steven Chu, the U.S. secretary of energy and a Nobel laureate in physics, has embraced the concept. He has cited a study from the Lawrence Berkeley National Laboratory that showed that increasing the reflectivity of roofs and pavements could radically reduce energy use, yielding billions of tons of greenhouse gas reductions.[88] Chu has initiated a series of "cool roof" projects under the aegis of the Department of Energy. California has required white for flat roofs for several years and "cool colored" surfaces for sloped roofs since July of 2009. It is also no surprise that white or light-colored cars use less air conditioning.

There has been a growing movement for some time, particularly in Europe, to build green roofs and walls. Vegetation on building surfaces not only looks good but has a number of significant practical aspects. The manifest benefits of green roofs and walls include excellent insulation to keep buildings cooler in summer and warmer in winter; an efficient way to moderate storm-water impacts, and to more easily collect the water for use within the building; a way for urban gardeners to plant vegetables, herbs, and flowers, and to teach horticulture; and the proven positive psychological effect of "biophilic" design for tenants.

There is nothing but upside in green building. A study commissioned by the USGBC had some impressive numbers among its conclusions.[89] From 2000 to 2008, the green construction market in the United States alone had

· generated $173 billion in GDP;
· supported over 2.4 million jobs;
· provided $123 billion in labor earnings.

From 2009 to 2013, green construction will

· generate an additional $554 billion in GDP;
· support over 7.9 million jobs;
· provide $396 billion in labor earnings.

Builders, developers, architects, buyers, renters, real estate agents, and planners are all becoming sold on the tremendous benefits of green building.

Green IT

Among the more aggressive proponents of clean tech are the folks in the industries that came into being because of high tech: Silicon Valley. It's no coincidence that so much of the venture capital and project finance, intense research and development, and deployment of technology is coming from this sector. Google, IBM, Cisco, and Microsoft are among the companies that are spearheading initiatives in the smart grid, renewables, energy efficiency, and, more and more, green IT.

The Climate Savers Computing Initiative, for instance, is a nonprofit consortium of seven hundred high-tech companies and conservation organizations that are seeking to significantly reduce the greenhouse gas footprint of computing operations all over the world. They set an ambitious goal of reducing worldwide carbon dioxide emissions by fifty-four million tons per year, equivalent to the annual output of eleven million cars. As of 2010, they had realized about two-thirds of that goal.[90] They have done this through increasing the energy efficiency of their equipment and encouraging best practices in power management throughout the industry.

The computer companies are also rolling out all manner of software and hardware to help manufacturers manage their environmental footprints, calculate the GHGs embedded in construction materials, expand the role of telecommuting and teleconferencing, and manage their supply chains. IBM even has a Sustainable Supplier Information Management Consulting business.[91] In fact, IBM's "Smarter Planet" initiative incorporates a broad and deep range of products and services, including some terrific information resources.[92]

The Future of Clean Tech

The world is in a transition to cleaner and smarter consumption and production. This is being driven by a number of important factors, including the looming threat of climate change, the rapid development of the technology, and the favorable economics, among other things. Sustainable development is no longer a dream. It's fast becoming a global reality.

breakthroughs

FROM RIO TO CANCÚN, AND BEYOND

Nopenhagen?

The popular view on the UN's talks on climate change in Copenhagen in December of 2009 was that they were a failure. There were scenes of protesters in the streets getting pummeled by the local police, while journalists and delegates queued up outside the meeting hall for hours in the freezing cold waiting to get in. People have the picture that inside the Bella Center various fractious blocs of small nations were walking out at key moments and in the backrooms the G7 nations in the persons of Barack Obama, Angela Merkel, Nicholas Sarkozy, Gordon Brown, and some others were betraying the conference altogether.

Copenhagen was the center of the known universe in December 2009. It was the site of the Fifteenth Conference of the Parties (COP 15) to the United Nations Framework Convention on Climate Change (UNFCCC), negotiated and agreed to by nearly all the world's nations in June of 1992 in Rio.* In the background to Copenhagen was the media-hyped controversy of "Climategate," which served, in some minds, to diminish the importance of the COP 15 delegates' work and indeed, for some others, to color the meeting as a folly.

More important, however, was the expectation that had built up over the course of two years since COP 13 in Bali. In Bali, a "roadmap" had been finalized that pointed to Copenhagen as the time and place where negotiations would culminate in a treaty to replace the Kyoto Protocol, agreed in 1997 and set to expire in 2012. The UN secretary-general, Ban Ki-moon, had continually insisted that the world needed to "seal the deal" in Copenhagen.

---

* The Persian Gulf oil states, plus Turkey and Syria, and some other small countries did not sign in Rio, but did subsequently.

Environmental organizations were pushing for a robust agreement. Business leaders and heads of state expected, for much of 2009, to come away with a new treaty. The media coverage was intense. Think tanks, policy centers in national governments, and international governmental organizations, NGOs, and multinational corporations generated a high volume of position papers on issues from adaptation to monitoring to finance to technology transfer.

Ban Ki-moon convened a "Summit on Climate Change" around the annual reopening of the UN General Assembly in New York in September. Over a hundred world leaders took part. Ban Ki-moon "sensed a keen willingness by every leader to contribute to the successful conclusion of negotiations in Copenhagen. They also expressed readiness to commit their nations to reach an effective agreement that is fully subscribed to, and acted upon, by all."[1] Brave words.

The wind started to come out of the sails soon thereafter. Although more meetings continued to underscore the progress being made and the good intentions of the major players involved, the sense began to grow that Copenhagen would not produce a treaty that the nations would embrace. There was considerable speculation in the fall as to whether Barack Obama would attend. The pace became frenetic in the weeks and days leading up to the meetings. Yvo de Boer, the executive secretary of the UNFCCC, indicated that although a final agreement might not be concluded, important aspects of a treaty could be negotiated, and some of the fine print could be filled in after the conference. This became the party line.

Whatever the headwinds, 10,500 delegates and 13,500 observers converged on Copenhagen. The media was represented by more than 3,000 souls. By the end, COP 15 had been attended by 120 heads of state and government.[2]

After twelve days that climaxed in some very dramatic behind-the-scenes negotiations, confrontations, and, in the end, some palpable failures of will, the conference produced the Copenhagen Accord. There was no formal vote to adopt the accord, but the delegates "took note" of it, giving it something of an official seal of approval. Some wags characterized the conference as Nopenhagen. One *Los Angeles Times* headline read: "Climate Summit Ends with Major Questions: 'Breakthrough' or 'Cop-out'?"[3] Greenpeace "strongly condemned the arrogance of the heads of state from the world's most powerful countries for presenting a 'take it or leave' deal to the Copenhagen Climate Summit."[4]

There were recriminations from high government officials. For instance, Ed Miliband, then the UK minister for climate (and now the leader of the Labour Party), said that China had prevented a more ambitious deal being signed. "We did not get an agreement on 50 per cent reductions in global emissions by 2050 or on 80 per cent reductions by developed countries. Both were vetoed by China, despite the support of a coalition of developed and the vast majority of developing countries."[5] The Chinese foreign ministry shot back: "The statements from certain British politicians are plainly a political scheme. Their objective is to shirk the obligations of developed countries to their developing counterparts and create discord among developing countries, but the attempt is doomed to fail."[6]

The world saw chaos at the highest levels. The implication was that the effort to confront the climate crisis had spun off its axis and was careering into space.

The Climate Continuum

Although Copenhagen did not live up to its prior billing, it produced some significant agreements. But critical to an understanding of how the world has been addressing climate change is the recognition that Copenhagen was but a station along the line. It was not, by any stretch of the imagination, the beginning or the end of the line.

The science, as we have seen, generated much of the early push on looking at and addressing warming and reducing greenhouse gases. The environmental consciousness, though, the sense that complex natural systems were being altered and badly impacted, goes back to the time of Rachel Carson's seminal 1962 book on the damage to the environment from pesticides, *Silent Spring*, and the rise of the modern environmental movement in the industrial economies in the 1960s, culminating in the first Earth Day, on April 22, 1970, and the ensuing flood of landmark environmental laws in the United States. The first vivid shots of our planet from space — NASA's "earthrise" photos from *Apollo 8* at Christmas of 1968 — resonated all over the world. The blue-green miracle of Planet Earth was both intensely beautiful as seen from space and much more easily understood as fragile.

The United States created its Environmental Protection Agency in early

1971, and other national and subnational jurisdictions around the world created special agencies to deal with pollution, often cobbling together various existing bureaucracies that had been working under the aegis of health, parks, and conservation authorities of various kinds. The UN hosted a "Conference on the Human Environment" in 1972 in Stockholm, which gave rise to the United Nations Environment Program (UNEP). International treaties ensued: these included the Convention on International Trade in Endangered Species (CITES) in 1973, the European Convention on Long Range Transboundary Air Pollution (acid rain) and the Bonn Convention on Migratory Species, both in 1979, and the United Nations Law of the Sea Convention in 1982.

The gathering of energy and momentum that had been building up in the 1960s had exploded into action, and, perhaps not surprisingly, enormous progress was made on water and air pollution, solid and hazardous waste management, and land conservation.

Philip Shabecoff's splendid history of the environmental movements, *A Fierce Green Fire*, describes it this way for America, but his description is applicable throughout much of the developed world: "The interest and anger displayed by millions of Americans had caused the politicians, the news media, the universities and other power centers to pay heed. With each passing year the environmental impulse became more deeply enmeshed in the nation's institutions, its laws, and its daily life."[7]

Attention to the environment was a new global value. At the time of the Stockholm Conference in 1972, only eleven nations had an environmental protection agency. By the time of the second UN Conference on the Human Environment ten years later, 106 countries did, more than half of them in the developing world.

At the second conference, nations reaffirmed the goals of environmental protection and acknowledged the progress that had been made. However, the conference called for a much greater focus on the existing problems and recognized new dangers:

Changes in the atmosphere—such as those in the ozone layer, the increasing concentration of carbon dioxide, and acid rain—pollution of the seas and inland waters, careless use and disposal of hazardous substances and the extinction of animal and plant species constitute further grave threats to the human environment.[8]

The robust science on stratospheric ozone depletion led the world to work toward the 1987 Montreal Protocol, an addendum to the 1985 Vienna Convention for the Protection of the Ozone Layer. That the convention and the protocol were so successful emboldened many in the international community of environmentalists to begin to think seriously about other big-ticket treaties, including one to address climate change.

Meanwhile, another sprout was beginning to grow, namely sustainable development. In early 1987, the Report of the World Commission on Environment and Development, titled *Our Common Future*, was published, propelling the term "sustainable development" into the public discourse. "In essence, sustainable development is a process of change in which the exploitation of resources, the direction of investments, the orientation of technological development . . . and institutional change are all in harmony and enhance both current and future potential to meet human needs and aspirations."[9]

A million copies of *Our Common Future* were printed, in thirty languages. The language of sustainable development, melding the critical message of environmental protection and natural resource conservation with that of raising people out of poverty, became a lingua franca across the world.

The Earth Summit

The Earth Summit in Rio in 1992 addressed a number of issues, climate change among them. Rio also produced a convention on biological diversity, an agreement on principles of forest management, and "Agenda 21," a nine-hundred-page action plan to protect the earth's natural environment. A Commission on Sustainable Development (CSD) was subsequently created by the UN to oversee this work. The CSD is served by the Division for Sustainable Development as its secretariat. Other UN agencies such as the UN Development Program (UNDP) and the UN Environment Program also carry this work forward.

William K. Reilly was the EPA administrator under President George H. W. Bush. Reilly had been a leading environmentalist for two decades prior to his appointment at EPA. He led the American delegation to Rio. He sees the Earth Summit as a key moment in environmental history and its decisions and documents as fundamental:

Agenda 21 basically are the governance concepts that have been universally acknowledged as desirable by the world community. We concretized understandings about environmental impact assessments, about areas of critical concern, about the fundamental importance of reconciling the ecology with the economy, and the kinds of structures that would be required and necessary to do that, including freedom of information and the opportunity for public involvement, respect for nongovernmental organizations.[10]

Rio was also "the largest international diplomatic conference ever held."[11] There were 179 nations represented — with 118 heads of state, eight thousand journalists, and representatives from over seven thousand NGOs.

Soon after Rio, Reilly wrote:

From a global perspective, the Earth Summit marked the arrival of environmental concerns on the international stage as a major new consideration in foreign policy. The presence in Rio of foreign ministers, prime ministers, development ministers and presidents made the point that environmental questions must be accommodated in decisions and policies affecting trade, energy, agriculture, and economic development.[12]

The Earth Summit, according to Reilly, was particularly important for developing nations. "Rio forced the political leadership of developing countries to learn the game. The reality is that environment had not typically been a high priority in the developing world." But, Reilly asserts, when world leaders and top ministers go to international conferences like Rio, they "certainly have to learn about the vulnerabilities of their own countries' policies and the trends of their countries which are working against what the rest of the world considers an appropriate sensitivity for nature."[13]

Maurice Strong, the prominent Canadian business leader, was the secretary-general of the Earth Summit. Going back to the early 1970s, Strong had been a driving force in creating momentum for action on the global environment. He was the principal organizer for the Stockholm Conference in 1972 and its secretary-general, became the first director of the United Nations Environment Program (UNEP), created in Stockholm, and was later one of the commissioners for the World Commission on Environment and Development that published *Our Common Future*. While at UNEP, he took steps to put climate change on the international agenda. Of Rio, Strong said it ignited "a wildfire of interest and support."[14]

The Framework Convention

The UN Framework Convention on Climate Change (UNFCCC) was perhaps the most important offspring of the Earth Summit. President George H. W. Bush, along with scores of other heads of state, signed the treaty in Rio. Within a few months' times, the convention was unanimously ratified by the U.S. Senate. Today there are over 190 parties to the convention.

The UNFCCC is clear in its objectives. It seeks "stabilization of greenhouse gas concentrations in the atmosphere at a level that would prevent dangerous anthropogenic interference with the climate system." Further, the parties to the convention "should protect the climate system for the benefit of present and future generations of humankind, on the basis of equity and in accordance with their common but differentiated responsibilities and respective capabilities."[15]

The phrase "common but differentiated responsibilities" is critical. It means that the developed economies of Europe, the United States, Japan, and others have the responsibility to take the lead in mitigating climate change and providing support for adapting to its impacts, including helping with finance and technical support to developing economies. This clause was included because the developed world has been the source, historically, of most of the greenhouse gases that are causing the problems, and because they have economic and technical capabilities that developing nations don't.

Like the Intergovernmental Panel on Climate Change, the UNFCCC has served as a key focus for all the stakeholders, from world leaders to indigenous peoples, from environmental activists to multinational corporate executives. The annual Conferences of the Parties — COP 17 having taken place in Durban, South Africa, in December 2011 — have been the focal point for much of the world's attention. However, there is a massive body of ongoing work performed by the UNFCCC secretariat, two subsidiary bodies (for scientific and technological advice and for implementation), plus a few expert groups helping countries draw up national action plans, advising countries on adaptation, and helping to facilitate technology transfer to less-advanced nations.

The secretariat has a staff of 350, led by an executive secretary, Christina Figueres, a Costa Rican well traveled in climate and energy circles, particularly in Latin America. Much of the work, however, is done by the 192 parties to the convention and 184 parties to the Kyoto Protocol.

Kyoto

The Kyoto Protocol is the most famous component of the framework convention. Aside from accepting the abundant science showing the stark reality of climate change, its anthropogenic origins, and the compelling need to confront it, one of the principal themes of the convention, as we have seen, is the idea of "common but differentiated responsibilities." It is stipulated by all the parties that because the industrial economies of the developed world were responsible in the first place for the large and growing stock of greenhouse gases that have been forcing the climate, then it should be the principal responsibility of these nations to provide the resources to address the problem, to "take the lead in combating climate change and the adverse effects thereof."[16]

The parties were grouped into several classes. Annex I parties are the developed countries, including the EU in its own right. There are forty-two Annex I parties in total. These are further divided into Annex II countries and those with economies in transition (EIT). The EIT countries are former Soviet republics and Soviet bloc countries. The non–Annex I parties are those in the developing world. The Annex II countries have the greatest responsibility and the non–Annex I the least.

The first Conference of the Parties after Rio took place in Berlin in 1995. The "Berlin Mandate" called for the institution of quantitative targets and timelines for emission reductions. It was at COP 3, in Kyoto in 1997, that these targets and timelines were adopted. It was not, however, until February 2005 that the Kyoto Protocol entered into force. The protocol required "enough countries to encompass 55% of the carbon dioxide emissions of the Annex I countries" to sign on before it could enter into being. It required the ratification of Russia to pass the threshold.

There was a furor in the United States over the mere idea of the Kyoto

Protocol. Six months before the protocol was even finalized, the U.S. Senate sent a fairly unmistakable shot across the bow: A resolution offered by Robert Byrd and Chuck Hagel expressed the sense of the Senate that the United States should not be a party to any "new commitments to limit or reduce greenhouse gas emissions for the Annex I Parties, unless the protocol or other agreement also mandates new specific scheduled commitments to limit or reduce greenhouse gas emissions for Developing Country Parties within the same compliance period."[17] The resolution passed 95-0. Even though Vice President Al Gore had had a pivotal role in the negotiations that produced the Kyoto Protocol, the Clinton administration, seeing the handwriting on the wall from the Byrd-Hagel resolution, did not offer it for ratification.

After Clinton and Gore left office at the beginning of 2001, the conservative government led by George W. Bush and Dick Cheney vocally and visibly stayed out of the protocol. The final blow to American participation in the Kyoto process was the announcement in 2001 by President Bush, with Cheney at his side, that the protocol "was fatally flawed in fundamental ways."[18] He also called the science into question.

The immediate and spontaneous worldwide reaction to Bush's announcement was negative. Protestors spilled into the streets in many world capitals. Bush had already gone back on his campaign pledge to seek mandatory economy-wide greenhouse gas reductions, and the announcement on Kyoto entirely shut the door on any progress that the American government might make during his presidency. In fact, there was a concerted effort to suppress climate scientists and to assiduously promote the agenda of the fossil fuel industries.

Moving On

As unfortunate and counterproductive as the Bush-era opposition to progress on climate and clean energy was, there was a prevailing sense in much of the rest of the world that the battle had been irrevocably joined to confront the climate crisis and that the world would proceed with or without the participation of the world's largest economy and biggest emitter of greenhouse gases.

The Third Assessment Report of the Intergovernmental Panel on Climate Change came out in 2001. It concluded, among other things, that "there is new and stronger evidence that most of the warming observed over the last 50 years is attributable to human activities."[19] Even the U.S. Global Change Research Project concluded in 2002 that "continuing growth in greenhouse gas emissions is likely to lead to annual-average warming over the United States that could be as much as several degrees Celsius (roughly 3–9 degrees Fahrenheit) during the 21st century. In addition, both precipitation and evaporation are projected to increase, and occurrences of unusual warmth and extreme wet and dry conditions are expected to become more frequent."[20]

In late August of 2002, the tenth anniversary of Rio, delegates from over one hundred countries met in Johannesburg for the World Summit on Sustainable Development. In addition to the country representatives, an array of international governmental organizations and nearly eight thousand people from 925 NGOs were in attendance at the official events, with thousands more at side events. Although the summit did not produce an entity as important as the UN Framework Convention on Climate Change, the event was yet another milestone on the ever-busier road to sustainability.

One of the eight Millennium Development Goals (MDG) adopted by world leaders at the UN in 2000 is to "ensure environmental sustainability." To do this, the aim is to integrate sustainability into all countries' policies, reduce biodiversity loss, halve the population of the world without access to clean drinking water and basic sanitation, and to improve habitat for slum dwellers. This and the other MDGs are now part and parcel of the work done by the UN and all its agencies.

Environmental Regimes

In addition to the conferences and the creation of key international bodies such as the UNFCCC, international agreements, bilateral and multilateral, have grown in number, size, and scope over the past several decades. It is estimated that environmental concerns are at least a portion of over a thousand international agreements. Fully 230 are centered on the protection of the environment.[21] Among the core treaties are MARPOL (International Convention for the Prevention of Pollution from Ships), the Law of the Sea Convention, CITES (Convention on International Trade in Endangered Species of

Wild Fauna and Flora), the Montreal Protocol (to the Vienna Convention for the Protection of the Ozone Layer), the Basel Convention on the Control of Transboundary Movement of Hazardous Wastes and Their Disposal, the Convention on Biological Diversity, and the Stockholm Convention on Persistent Organic Pollutants.

We have touched on how the Montreal Protocol gave hope to those seeking an international regime on controlling the forces driving climate change. That such an agreement could be made, in a relatively short amount of time, emboldened those who felt that a treaty reducing greenhouse gases, admittedly a much greater challenge than simply reducing the production and use of ozone-depleting substances (ODS), nevertheless might be possible.

What was not understood at the time of the Montreal Protocol was the huge global warming potential of most of these ODSS. A scientific study published in 2007 noted the ozone-layer protecting agreement was responsible for a 50 percent cut in the amount of greenhouse warming that would have occurred by 2010 had these substances continued to accumulate unabated in earth's atmosphere. The study said it this way: "Climate protection already achieved by the Montreal Protocol alone is far larger than the reduction target of the first commitment period of the Kyoto Protocol."[22]

With the phasing out of the ozone-depleting chlorofluorocarbons and other ODSS, new chemicals were brought into production to serve as refrigerants and for other purposes. Hydrofluorocarbons (HFCs) are much less harmful to the stratospheric ozone layer, but they are nevertheless potent global warming agents and regulated under the Kyoto Protocol as such. Two other Kyoto gases, perfluorocarbons (PFCs) and sulfur hexafluoride (SF_6), will stay in the atmosphere for hundreds of years and have very high global warming potential (GWP) relative to carbon dioxide.* PFCs range from 7,390 to 12,200 times the power of CO_2, and SF_6 about 22,800.[23]

---

* The IPCC defines Global Warming Potential (GWP) as "an index, based upon radiative properties of well mixed greenhouse gases, measuring the radiative forcing of a unit mass of a given well mixed greenhouse gas in today's atmosphere integrated over a chosen time horizon, relative to that of carbon dioxide. The GWP represents the combined effect of the differing times these gases remain in the atmosphere and their relative effectiveness in absorbing outgoing thermal infrared radiation. The Kyoto Protocol is based on GWPs from pulse emissions over a 100-year time frame."

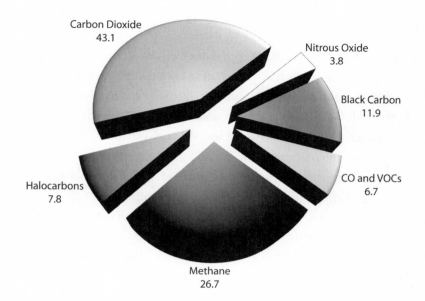

Carbon Dioxide
43.1

Nitrous Oxide
3.8

Black Carbon
11.9

CO and VOCs
6.7

Methane
26.7

Halocarbons
7.8

4.1 The sources of global warming (percentage of total). Graphic by author and Lorraine Simonello, based on data from Drew T. Shindell et al., "Improved Attribution of Climate Forcing to Emissions," *Science*, October 30, 2009, pp. 716–18

Although the greatest single agent of climate forcing is carbon dioxide, largely from the combustion of fossil fuels for power, industry, and transportation, the importance of these three "industrial gases" plus the two other Kyoto-regulated gases, nitrous oxide and methane, are very much recognized and acknowledged as a problem, and are being addressed.

Other Greenhouse Gases

Other pollutants have been seen as powerful climate-forcing agents as well. These include black carbon, more commonly known as soot, as well as ground-level (or tropospheric) ozone, carbon monoxide, and volatile organic compounds. NASA scientists, among others, assign a much higher importance to methane and black carbon as contributing to the overall problem of climate forcing than has previously been felt. Methane, an analysis of one key study concludes, is responsible for as much as 26 percent of the problem, and black carbon, 12 percent.[24]

We have seen that Kyoto didn't come into force until 2005 when Russia entered into the agreement. It is important to note that Kyoto was never meant to be the end of the process. Kyoto was in fact seen as an important next step along the path from Rio to a truly comprehensive and effective international approach to confronting climate change. What Kyoto did do was to lay some very important foundations for action and get things rolling on quantifiable emissions reductions—all in the context of sustainable development and the "common but differentiated responsibilities" of nations.

The developed economies—the Annex I countries in the terminology of the UNFCCC—assumed the greatest burden for reducing emissions. They agreed to kick start the process by getting their emissions down or at least holding them below a level that would otherwise occur in a business-as-usual scenario. The EU signed on to an aggregate 8 percent decline by 2012 from 1990 levels. The United States agreed in Kyoto to a 7 percent reduction, but, of course, the United States never ratified the treaty, and President George W. Bush pulled out of the Kyoto process altogether in 2001. Canada, Hungary, Japan, and Poland were in for 6 percent. New Zealand, Russia, and Ukraine said they would keep their emissions to 1990 levels. (Canada, unfortunately, has since pulled out of the Kyoto Protocol.[25])

The six greenhouse gases to be reduced, as we have seen, are carbon dioxide (CO_2), methane (CH_4), nitrous oxide (N_2O), hydrofluorocarbons (HFCs), perfluorocarbons (PFCs), and sulphur hexafluoride (SF_6). Because of the varying quantities of these GHGs and their differing global warming potential, for the purposes of standardization the common measure of their impact is measured as carbon dioxide equivalent, CO_2eq (or CO_2e).* Thus, for instance, the EU's responsibility under Kyoto is to reduce its total emissions, measured in CO_2eq, 8 percent from 1990 levels by 2012. The UNFCCC secretariat in Bonn administers a reporting system and a compliance system to ensure that parties are meeting their commitments. Should any Annex I party that ratified

---

* Carbon dioxide equivalent (CO_2eq) is defined by the Worldwatch Institute as "a unit of measurement used to compare the climate effects of all greenhouse gases to each other. CO_2eq is calculated by multiplying the quantity of a greenhouse gas by its global warming potential."

the protocol and agreed to emission reductions be found not to be meeting its commitments, either by having a flawed national system for monitoring and reporting its GHG output, or by failing to hit its assigned numbers, then the UNFCCC, working through a very deliberative process, may declare the party to be in noncompliance. It can require the submission of a plan to address the problems, and even declare the party to be ineligible to take part in the market-based mechanisms that are central to the workings of the protocol. It may also raise the required amount of the commitment to reduce emissions. So far, only Croatia has been found to be in noncompliance and had its eligibility to participate suspended. Romania and Ukraine have also been found to be in noncompliance but have retained the right to take part in a special trading program.[26]

The Flexible Mechanisms

Each country determines for itself how it will meet its target. A regulatory approach, in which strict caps are set for various sectors such as electrical utilities, is one approach. Establishing a "price on carbon" by imposing high taxes on fossil fuels is another way to get at the problem.

In support of the parties' national programs, the Kyoto Protocol offers three market-based "flexible mechanisms" to drive investment, to help parties meet their goals, and to channel money toward sustainable projects in the developing world. These programs are emissions trading, the Clean Development Mechanism (CDM), and Joint Implementation (JI).

Emissions trading has evolved into a vibrant and growing global "carbon market" that has taken some of the sting out of the process of getting GHGs down. The CDM similarly helps countries manage their quotas for reductions by allowing them to "offset" some of their emissions by buying Certified Emission Reduction (CER) credits generated by sustainable projects in the developing world. The third flexible mechanism is something of a hybrid. It allows Annex I countries to work with others in that category to create projects that reduce emissions. Russia is the host country for the majority of these projects, with Ukraine and other former Soviet bloc states involved as well. For example, Japan will receive credits equaling about three million tons of CO_2eq by virtue of helping to finance a project that recovers gas from a Russian oil field. The gas being recovered is a potent greenhouse gas.[27] We will look much more closely at all three of the flexible mechanisms in the next chapter when we treat the critical role of international environmental finance.

Europe's Leading Role

The Western European countries and the European Union itself have been motive forces in addressing climate change, as well as transitioning from fossil-fuel dependency and toward sustainability. The first international conference on climate change science took place in Austria in 1985, and its report was written largely by European researchers. The Germans wanted a binding commitment on climate as early as the Earth Summit in 1992 and have been in the vanguard ever since on renewable energy and on promoting a vigorous greenhouse gas scheme within Europe and through the UNFCCC. Top British government officials from both the Liberal and the Conservative parties have consistently pressed for action on climate change. Spain, Denmark, and Germany have been rapidly building out both their renewable energy generating capacity and their renewables manufacturing industries, with a growing international presence.

The European Union Emissions Trading System (EU ETS), a comprehensive "cap-and-trade" regime, was launched in 2005, in advance of and separate from the Kyoto system. When Kyoto came into force, the ETS incorporated the flexible mechanisms. Cap and trade is the innovative emissions reduction scheme that was first deployed under the 1990 Clean Air Act in the United States to reduce acid rain precursor pollutants like sulfur dioxide. It combines strict regulation of emissions with an ability to market and sell "credits" for achieved reductions. In part because of its considerable success in the United States, and also because leaders in European industry recognized the need for a market-based mechanism to mitigate the economic impacts of the cap, the Europeans embraced cap and trade for their far-reaching system to address greenhouse gases.

The EU ETS covers about eleven thousand power stations and industrial plants in thirty countries: the twenty-seven EU member states plus Iceland, Liechtenstein, and Norway. We will look more closely at the ETS in the next chapter on environmental finance.

At the beginning of 2008, the European Commission proposed an ambitious package that would first of all deepen and strengthen the ETS. The keystone of the EU policy was to be the "20/20/20" program. By the year 2020, the European Commission aimed for

- a reduction in EU greenhouse gas emissions of at least 20 percent below 1990 levels;
- 20 percent of EU energy consumption to come from renewable resources;
- a 20 percent reduction in primary energy use compared with projected levels, to be achieved by improving energy efficiency.

Among the 27 member states of the European Union, there is a disparity between the economies of the more developed EU-15 and those of the EU-12, the newer states, many of which are in the old Soviet bloc portion of Central and Eastern Europe. Their power and industry are heavily dependent on coal and so are coming along more slowly in lowering carbon dioxide output. Overall, though, total GHG decreased 15.5 percent in the EU from 1990 to 2010, while GDP grew in the same period by 41 percent.[28] While the waste, agriculture, power, and industrial sectors all managed healthy declines in GHGs, one sector, transportation (cars, trucks, marine shipping, and aviation), has had increased emissions.[29]

Notwithstanding the increase from transportation, progress has been steady, and, in July of 2010, France, Germany, and the UK jointly called for an increase in the GHG target from a 20 percent to 30 percent reduction.[30] On renewables, as we have seen, there have been some stunning breakthroughs in Europe. They are already more than halfway there to the 20 percent target they have set for themselves. For Europe, the sky is the limit for renewable energy. There are, of course, concomitant greenhouse gas reductions that go with renewable energy generation and other clean-tech deployments.

Beyond Europe: Pricing Carbon Down Under

Things have been advancing in Australia as well. In the fall of 2007, Australians threw out the government led by John Howard that for years had been dragging its feet on action on climate change. The first thing that Kevin

Rudd, the new prime minister, did was to sign the Kyoto Protocol. Australians had been experiencing the crush of drought and heat for years and were more than ready to confront the problem by 2007. Interestingly, Rudd suffered from his subsequent inability to push through a carbon trading law. Australians wanted *more* action. He fell from power within his party in 2010, to be replaced by Julia Gillard, and the Labor Party narrowly survived in national elections, forming an alliance with Independents and Greens. This coalition has pushed the government toward a stronger approach on climate and energy. "At the core of Australia's policy framework must be a price on carbon," a top executive of a major Australian power generator said at the time of the elections.[31] Australia moved steadily forward in that direction until, by late 2011, a greenhouse gas reduction scheme was voted into law by the federal legislature. The program will begin with a tax in 2012, then become a cap-and-trade regime by 2015. In instituting this key initiative to reduce greenhouse gases, Australia joined its antipodean neighbor New Zealand, which instituted its own Emissions Trading Scheme (NZ ETS) in 2008.

How U.S. Politics Come into Play

As important as these advances have been in Europe, Australia, and New Zealand, much of the hope for a successful effort to eventually turn back the tide of climate-changing emissions still depends on the United States. So, because of the critical role of the United States in any international negotiations on climate and energy, and, for that matter, on other key sustainability initiatives, it is necessary to keep U.S. domestic politics in mind. In a republic like the United States, key public policy in areas such as energy, trade, and international relations is usually determined by the composition of the national legislature.

In the midterm elections of November of 2006, the Democratic Party established a majority in the House of Representatives and won enough Senate seats to create a de facto majority there. The leadership and the committee makeup in both Houses reflected this new political reality. With these changes, it was possible to advance legislation in Congress that had hitherto been bottled up. With the grudging support of President George W. Bush, the

Energy Independence and Security Act (EISA) of 2007 jump-started a revived American push on clean energy. EISA set new standards for energy efficiency, particularly for federal facilities and vehicle fleets, boosted research and development on renewable energy technologies, and gave support to initiatives like the smart grid, green jobs training, and vehicle electrification.

Nancy Pelosi, the new Speaker of the House, created a special Select Committee on Energy Independence and Global Warming that focused a lot of attention on the issues through the hearing process.

In 2008, the Democrats gained even more seats. One particular quiet revolution ensued when staunch environmentalist Henry Waxman replaced John Dingell as chair of the all-important Energy and Commerce Committee in the House of Representatives. Dingell, although progressive on many issues, had been a key ally of the U.S. auto industry on Capitol Hill for years and had held up important environmental legislation along the way. Waxman then became the motive force behind a comprehensive climate and energy vehicle, the American Clean Energy and Security Act, which passed in the House, narrowly, in June 2009. ACES included a complex but, by most analyses, robust and fair economy-wide cap-and-trade program. The U.S. Senate, as has been well documented, failed to arrive at anything resembling a consensus on similar legislation, and so cap and trade died in the 111th Congress.

The fall of 2008 was an epically turbulent time with the collapse of Lehman Brothers and the ensuing panic about the fate of the world's financial systems. One key response in the United States was the Emergency Economic Stabilization Act of 2008, with the Troubled Asset Relief Program (TARP) at its core.

One of the little-recognized but nevertheless critical facets of the legislation was the extension of billions of dollars' worth of tax credits for the solar, wind, and other renewable industries. These industries had been, in a word, desperate for the extension of these programs. The TARP also included the creation of some new mechanisms such as support for plug-in hybrid electric vehicles.

The Obama Effect

Aside from the Democratic Party gains in 2008, there was one singularly dramatic political development with enormous significance for progress both in the United States and internationally on climate and energy: the election of Barack Obama to the White House. Obama ran on a strong environmental platform, including support for a far-reaching and vigorous approach to climate change.

Even before his inauguration, he signaled a breathtaking reversal from policies fostered by Bush and Cheney for eight years. His core transition team, for instance, included stalwart renewable energy advocates like Dan Reicher. At COP 14 in Poland in the late fall after the election, delegates were feeling an "Obama buzz." John Kerry, Bush's opponent in 2004 and the chairman of the Senate Foreign Relations Committee, was quoted at the meetings: "America is back. After eight years of obstruction and delay and denial, the United States is going to rejoin the world community in tackling this global challenge."[32]

Brian Urquhart, an extraordinary writer on international affairs and witness to history, put it this way:

> A fog of know-nothing ideology, anti-intellectualism, cronyism, incompetence, and cynicism has, for eight years, enveloped the executive branch of the United States government. America's role in the world and the policies that should shape and maintain it have been distorted by misguided decisions and by willful misinterpretations both of history and of current events. That fog is now being dispersed, and the vast intellectual and managerial resources of the United States are once again being mobilized.[33]

This applies, categorically, to climate change and energy policy.

Obama's cabinet choices were strong proponents of action on climate change: Hillary Clinton for the State Department; former Iowa governor Tom Vilsack at Agriculture; Colorado senator Ken Salazar at Interior; Steven Chu, a Nobel laureate in physics, at the Department of Energy; and, at EPA, Lisa Jackson, a chemical engineer by trade and an experienced regulator.

John Holdren, a physicist, Harvard professor, and director of the Woods Hole Research Center, was named Obama's top science adviser. Holdren is a passionate advocate of early and decisive action on warming. Jane Lubchenco was tapped to head the National Oceanic and Atmospheric Administration (NOAA), an agency that has critical importance in monitoring the environmental changes that are occurring. She, like Holdren and Chu, was an outspoken member of the scientific community on the issue of climate change.

Green Stimulus

The President-elect and his top officials, working with congressional leaders, were crafting an economic stimulus package. Given the parlous state of the economy in the United States, this was felt to be an important necessary step to be taken early in the new administration. Elsewhere in the world, similar large booster shots to the big economies were taking shape.

Leading economists identified the necessity of "greening" the stimulus packages that would be forthcoming from the world's leading economies. Lord Nicholas Stern, the lead author of the UK's pathbreaking study in 2006 on the economics of climate change, said in 2009 that "with billions about to be spent by governments on energy, buildings and transport, it is vital that these public investments do not lock us for many more decades into a costly and unsustainable high-carbon economy."[34]

The Obama administration understood this, and of the $787 billion to be made available through the American Recovery and Reinvestment Act, about $112 billion was to be devoted to green stimulus (figure 4.2).

These programs have, in fact, been a shot in the arm to clean-tech industries and the work force. One analysis from the White House Council of Economic Advisers estimates that 827,000 total job-years* in clean energy will have been supported by direct spending from agencies and tax provisions through the end of 2012.[35]

---

* A job-year is one person employed for one year.

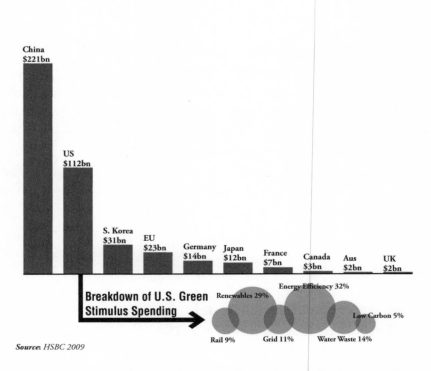

China
$221bn

US
$112bn

S. Korea
$31bn

EU
$23bn

Germany
$14bn

Japan
$12bn

France
$7bn

Canada
$3bn

Aus
$2bn

UK
$2bn

Breakdown of U.S. Green Stimulus Spending → Renewables 29%

Energy Efficiency 32%

Low Carbon 5%

Rail 9%   Grid 11%   Water Waste 14%

Source: HSBC 2009

4.2 Green stimulus spending by country (in billions of U.S. dollars). WRI. 2009. "Countdown to Copenhagen: U.S. Climate Actions," Washington, DC: World Resources Institute (available online: http://www.wri.org/stories/2009/04/countdown-copenhagen-us-climate-actions)

The American Auto Industry

One of the most important sectors of the U.S. economy has been and continues to be automotives. Much of the American auto industry had been on the ropes for years, but the economic near-meltdown in 2008 drove the Big Three — General Motors, Chrysler, and Ford — to the brink. The Obama administration lent considerable support to the automakers, allowing them to catch their breath, restructure, and return to health by 2010. At the same time, Obama, with his EPA and Department of Transportation teams, negotiated a deal in the spring of 2009 with the industry, key elected officials, the United Auto Workers, and environmental organizations to establish new standards for fuel economy and GHG emissions from cars and light trucks.

This deal marked another 180-degree turn from the Bush administration, which had refused to consider California's request to regulate GHG in vehicles — the first time in the nearly forty years of the Clean Air Act that the EPA had denied a waiver request from California. Because of the particularly difficult air pollution problems that the Golden State has experienced historically, mostly in Southern California, the Clean Air Act allows it to ask for different and more stringent ways of getting at the problems than the federal government may mandate. The Obama administration approach was not to do the job piecemeal, simply allowing California and other states that have usually piggybacked on California's tougher regulations to implement GHG controls on emissions by themselves. Instead, the idea, embraced by the industry, was to work toward new standards for cars and trucks being sold throughout the United States.

According to the White House, the new rules are "projected to save 1.8 billion barrels of oil over the life of the program with a fuel economy gain averaging more than 5 percent per year and a reduction of approximately 900 million metric tons in greenhouse gas emissions."[36] A year later, a similar deal on medium and heavy-duty trucks was announced. At the same time, the president directed the Department of Energy to increase its support for advanced vehicles, including electric vehicles.[37]

Big-Ticket Federal Government Initiatives

In June of 2009, Obama announced new standards for energy efficiency for appliances and lighting, leading to massive reductions in power use, eliminating the need for the juice from as many as fourteen coal-fired power plants.[38] In October of that year, Obama signed an executive order setting sustainability goals for government buildings and fleets, which were meant to lead to a reduction in greenhouse gas emissions of 28 percent by the federal government by 2020. Will that have a significant overall impact? Consider this: The government has space in nearly half a million buildings, operates more than six hundred thousand vehicles, employs more than 1.8 million civilians, and purchases more than $500 billion per year in goods and services.[39]

The Department of Defense (DOD) is moving vigorously forward in deploying renewable energy at its bases and alternative fuels for its fleets, very much including aircraft. The Air Force has certified nearly all of its aircraft fit to run on biofuels as of 2011.[40] One very senior DOD official is particularly gung ho, as Elisabeth Rosenthal of the *New York Times* reports: "'There are a lot of profound reasons for doing this, but for us at the core it's practical,' said Ray Mabus, the Navy secretary and a former ambassador to Saudi Arabia, who has said he wants 50 percent of the power for the Navy and Marines to come from renewable energy sources by 2020. That figure includes energy for bases as well as fuel for cars and ships."[41] Fifty percent by 2020! *New York Times* columnist Thomas Friedman celebrated Mabus's goals: "If the Navy really uses its buying power when buying power, and setting building efficiency standards, it alone could expand the green energy market in a decisive way."[42] The DOD is on track to be spending $10 billion a year by 2030 on renewables.[43] As one leading member of the U.S. Senate in the 1950s and '60s, Everett Dirksen, is alleged to have said: "A billion here, a billion there, and pretty soon you're talking real money." Indeed.

On the science side, the U.S. Global Change Research Program (USGCRP) looks at all aspects of federal research on climate and other environmental changes, and their impacts. Thirteen federal entities are involved, including DOD, DOE, and EPA, plus the departments of Agriculture, Commerce, Health and Human Services, and Interior, and the National Science Foundation, the Smithsonian, and the U.S. Agency for International Development, among others. Of course, these individual agencies themselves have vigorous science and technology research and development programs, staffed by top experts in their fields. The fiscal year 2010 budget that was enacted for the work of all the agencies overseen by the USGCRP in six program areas in climate science was $2.178 billion.[44]

The Department of the Interior has a key role to play in leasing public lands for renewable energy development. In October 2010, Secretary Salazar made a series of announcements on granting the use of public land for several pathbreaking installations of utility-scale solar power. He also signed the lease that same month for the first offshore wind farm in the United States, the controversial Cape Wind. Salazar, working hard to streamline the process for these sorts of permissions, said at the time, "Responsibly developing this clean, renewable, domestic resource will stimulate investment in cutting-edge technology, create good, solid jobs for American workers, and promote our nation's competitiveness, security, and prosperity."[45] The United States has a lot of catching up to do on offshore wind, but there's a huge upside that will certainly be facilitated by Interior's efforts.

EPA's Voice

The Clean Air Act of 1970 was a milestone in environmental protection for the United States. The Clean Air Act requires the EPA to regulate air pollutants that are found to be an "endangerment" to public health. A petition filed by nineteen environmental and renewable energy industry organizations in 1999 sought the regulation of greenhouse gases by the EPA. The Bush administration maintained that GHGs were not subject to regulation under the law, but the Supreme Court ruled in April of 2007 that not only were GHGs air pollutants but that the EPA was mandated to determine whether or not they constituted a danger. Early in the Obama administration, the EPA announced its initial finding of endangerment, and in December of 2009, EPA administrator Lisa Jackson signed a finding that "six key well-mixed greenhouse gases . . . in the atmosphere threaten the public health and welfare of current and future generations."[46] This finding applied to light-duty vehicles, but set the stage for further regulation of greenhouse gases.

The EPA's approach to climate change has, since 2009, been comprehensive. It has coordinated efforts on science with NASA, NOAA, and other agencies; has negotiated the extraordinary and historical agreement with the auto industry to regulate GHGs described above; has created a greenhouse gas registry program; and has been moving along at a measured but steady pace to rein in emissions that are exacerbating climate forcing. We will see in Chapter 6 how the EPA's approach, absent any new legal authority from Congress — and, in fact, in the teeth of virulent opposition from some quarters — is pushing the edge of the envelope.

American Global Policy

It is important to realize the Obama administration has seized the initiative in the international arena as well as in the domestic. One very strong voice on climate change has been the secretary of state, Hillary Clinton. In an op-ed at the time of the Copenhagen meetings, she wrote: "Our world is on an unsustainable path that threatens not only our environment, but our economies and our security. It is time to launch a broad operational accord on climate change that will set us on a new course." Another key presence has been the special envoy for climate change, Todd Stern. Stern came in with extensive experience on climate change and international relations from Bill Clinton's administration and from the nonprofit sector.

One enterprise the Obama administration has launched is the Major Economies Forum on Energy and Climate. In March of 2009, he invited the leaders of sixteen major economies and the UN secretary-general to send representatives to participate in creating this body.[47] High-level meetings have been taking place since then with an emphasis on developing "transformational low-carbon, climate-friendly technologies" with ten task forces led by individual or pairs of countries looking at everything from advanced vehicles (Canada) to marine energy (France) to the smart grid (Italy and Korea).[48]

In September 2009, Obama hosted the G20 economic summit in Pittsburgh. He formally proposed there to phase out the nearly $600 billion in fossil fuel subsidies spent annually across the world. The G20 Summit embraced that commitment, saying it would "improve energy security, encourage investment in clean energy sources, promote green growth, free-up resources to use for pressing social needs such as health, food security, and environmental protection."[49] Fatih Birol, chief economist at the International Energy Agency, has said: "I see fossil fuel subsidies as the appendicitis of the global energy system which needs to be removed for a healthy, sustainable development future."[50]

Bilateral agreements on energy and climate are critical components of solving these thorny global problems. The United States and India, for instance, have strong programs under way, including the U.S.-India Partnership to Advance Clean Energy, through which a Joint Clean Energy Research and Development Center will be established that will mobilize up to $100 million in public and private-sector funding over five years. Another example of cooperation is in the Overseas Private Investment Corporation, providing $100 million in financing for the $300 million Global Environment Fund (GEF) South Asia Energy Fund. Similarly, Indonesia and the United States are furthering the $119 million SOLUSI partnership, which has programs in tropical forest conservation, marine conservation, and clean energy development, among others. When President Obama was in these two countries in November 2010, he made a point of emphasizing the importance of these sorts of initiatives.

Bilateral and Multilateral Climate and Energy Innovation

Many other countries are engaged in developing relationships to foster clean tech and natural resource conservation. Norway, for instance, a country well endowed by its fossil fuel riches, and a well-managed sovereign wealth fund into the bargain, has become a leader in promoting and financing rain-forest conservation. In May 2010, Norway pledged $1 billion to Indonesia for avoiding deforestation. Indonesia, because of its rampant conversion of peatlands, wetlands, and rainforest to palm oil plantations, is the third-largest contributor to climate forcing among the world's nations. (China and the United States are one and two, with Brazil, because of its deforestation, at number four.) Indonesia has made excellent progress, on paper, of curtailing its devastating practices, but the scope of the problem and a longtime habit of corruption are impediments to a robust program. Norway, for its part, has no intention of pouring its money down a sinkhole, as has happened so many times in international development programs, far too often routinely. It is working with Indonesia to develop guidelines and safeguards that will help to ensure that deforestation will, in fact, be radically reduced.

Another bilateral program is the Sumatra Forest Carbon Partnership between Indonesia and Australia, part of Australia's $273 million International Forest Carbon Initiative.

Norway is also a billion-dollar donor to Brazil's Amazon Fund. The fund was set up in August 2008 to receive donations from nations, multilateral organizations, companies, nonprofits, and individuals "to prevent, monitor and combat deforestation, as well as to promote the preservation and sustainable use of forests in the Amazon Biome."[51] It aims to raise $21 billion over the next decade. Brazil, under President Luiz Inácio Lula da Silva from 2003 through 2010, made significant progress on turning back the tide of rainforest destruction. Brazil is itself working bilaterally with Mozambique on "South-South REDD: A Brazil-Mozambique Initiative." We will delve much deeper into the centrally important Reducing Emissions from Deforestation and Degradation (REDD+) program in Chapter 7.

The Heinrich Böll Foundation, affiliated with the German Green Party, manages Climate Funds Update, which lists twenty-two international programs working on the mitigation of and adaptation to climate change. This regularly updated inventory accounts for $32 billion in pledges from donor countries as of mid-2011. Japan is the world leader by far, with over half of the total pledged.[52]

There is clearly a tremendous commitment and a concomitant amount of funding activity from nations, multilateral organizations like the World Bank, foundations, and nonprofits. We will look more thoroughly at these and at corporate spending in the next chapter, as well as in the chapter on sustainable development.

Returning to the subject of bilateral cooperation on climate change, sustainability, and energy, it is important to look at the two biggest contributors to climate forcing in the world: China and America. These two countries represent 40 percent of the total. Each country understands that the eyes of the world, including large numbers of their own citizens, are on them. The governments in each country also understand the stakes. China's reputation was tarnished in Copenhagen by its overt and covert obstructionism. The United States, although it performed well in helping to stitch together the final compromise agreement, still came out looking bad, largely as a consequence of so much being expected of the world's largest economy and most-powerful democracy.

One key way to address the problems that each country faces is for them to cooperate on initiatives like clean energy, energy efficiency, and electric vehicles. One initiative that is under way is the creation of the U.S.-China Clean Energy Research Center. In November 2009, when President Obama was in China, the two countries issued a statement that outlined a number of these types of programs and acknowledged "that climate change is one of the greatest challenges of our time. The two sides maintain that a vigorous response is necessary and that international cooperation is indispensable in responding to this challenge."[53]

Copenhagen to Cancún to Durban

It is important to realize that in Copenhagen, in December 2009, China, and other major developing industrial powers like India, made it plain that they are willing and able to reduce their emissions. A leading climate activist for the highly regarded international environmental organization, the

Natural Resources Defense Council (NRDC), put it this way: "The Accord is a breakthrough because, for the first time, all major economies, including China, India, and Brazil, as well as the United States, Russia, Japan and the EU, have made commitments to curb global warming pollution and report on their actions and emissions in a transparent fashion, subject to 'international consultations and analysis.'"[54] President Obama put it even more succinctly: "For the first time in history all major economies have come together to accept their responsibility to take action to confront the threat of climate change."[55]

Copenhagen, messy as it was, produced a very useful outcome. The accord required countries to make a submission accounting for actions they were going to take to reduce emissions in the near term and through mid-century. Within three months of the meetings, seventy-three countries had formally communicated their targets. The Annex I parties—the developed economies—including the EU and its member nations, the United States, Japan, Russia, Canada, and Australia, were on board. The non–Annex I nations—the developing economies—China, India, Brazil, South Africa, and South Korea, among others, also all submitted plans by the deadline.

China's submission, for instance, said: "China will endeavor to lower its carbon dioxide emissions per unit of GDP by 40–45 percent by 2020 compared to the 2005 level, increase the share of non-fossil fuels in primary energy consumption to around 15 percent by 2020 and increase forest coverage by 40 million hectares and forest stock volume by 1.3 billion cubic meters by 2020 from the 2005 levels."[56] The United States pledged to reduce economy-wide emissions 17 percent from 2005 levels by 2020,[57] while the EU and its member states committed to a 20 percent reduction from 1990 levels by 2020.[58]

In the year between COP 15 and 16, which took place in Cancún in December 2010, there were many meetings to try to crystallize some of the tacit agreements made in Copenhagen and on which the UN, the G20, many interested NGOs, and others have been working for years, namely finance, forests, adaptation, and technology transfer.

One of the thorniest, most persistent concerns about the international process begun in Rio in 1992 has been the distrust between developing and developed nations. Coming out of Cancún, though, there was a renewed spirit of cooperation. UNFCCC executive secretary Christina Figueres said, "The beacon of hope has been reignited and faith in the multilateral climate change process to deliver results has been restored."[59]

Beyond a restoration of faith, there were concrete gains. For one thing, the Cancún Agreements reiterated the commitments on emissions from all the world's major economies. Even more, they provide for, in the words of Hillary Clinton, "a system of transparency, with substantial detail . . . which will provide confidence that a country's pledges are being carried out."[60]

The whole process of reducing and indeed reversing deforestation and forest degradation was advanced. In the UN system and beyond, REDD — Reducing Emissions from Deforestation and Forest Degradation — has served to empower developing nations to curtail their most destructive forest management practices, thus reducing the massive greenhouse gas burden of these activities. It is now understood that the next stage, REDD+, is not only about reducing emissions but halting and reversing forest loss. It is also now accepted that the responsibility for implementing REDD+ lies not only with the countries like Indonesia and Brazil where forests have been decimated, but with the countries that are driving the demand for the products that are causing the deforestation, like palm oil, soy, cattle, and timber.

On climate finance, a process to design a UN Green Climate Fund was established. To underwrite the fund, $30 billion in "fast start" finance has been pledged from developed nations to be used through 2012. As stipulated prior to Copenhagen, the goal is to provide $100 billion for developing nations annually by 2020 for mitigation and adaptation.

Technology transfer has been a goal of the parties since Rio. One of the agreements in Cancún was to create a new Technology Mechanism, which will have two components: the Technology Executive Committee and the Climate Technology Center and Network. The Technology Mechanism further elevates the process of support for clean tech within the UN process.

On adaptation, there is a new Cancún Adaptation Framework, which identifies a set of high-priority actions, including the building out of adaptation plans, improving research and increasing the quality of adaptation technology, and performing critical assessments of vulnerability and financial need.

It has been a long road from Rio. The Seventeenth Conference of the Parties took place in Durban, South Africa, at the end of 2011. Although there were no major breakthroughs, progress continued on all fronts. There have been, to be sure, many advances since 1992, but most observers think that the progress has not been commensurate with the scope and intensity of the problem. To say that climate change has been adequately addressed in the nearly twenty years since the Earth Summit would be foolish. It is fair to say, however, that progress that might not have been predicted even ten years ago is taking place now and that the pace of much of that progress is accelerating, and focus on the solutions to the climate crisis is intensifying.

follow the money

ENVIRONMENTAL FINANCE AND GREEN BUSINESS

The Rise of Waxman-Markey

On June 26, 2009, the U.S. House of Representatives passed the American Clean Energy and Security Act by a vote of 219 to 212. Universally known as Waxman-Markey, named for its two authors, the bill was a long time in the making and fulfilled the expectations of the majority of, if not all, energy and climate activists, as well as a very large segment of the business and finance communities. There were 168 Republicans voting no and 44 Democrats. Only 8 Republicans voted yes.

Those voting no were almost all from the suburban and rural South, Midwest, and West. Ironically, these are the three broad areas of the United States the hardest hit so far by climate change impacts and with the greatest potential for considerable further economic and ecological damage: the Gulf Coast region of the South by powerful hurricanes, the South generally by drought, the Midwest predicted to be subjected to persistent and intensifying heat and drought in coming years, and the West undergoing long-term drought and a devastating beetle infestation that has been decimating coniferous forests. According to the U.S. Global Change Research Program, changes in the climate "are already affecting water, energy, transportation, agriculture, ecosystems, and health. These impacts are different from region to region and will grow under projected climate change."[1]

The number, as well as the political and geographic distribution, of those voting no was an ill boding for complementary action in the U.S. Senate.

The political landscape for climate and energy changed considerably in the United States in 2006 when the Democrats, riding a wave of discontent over the policies of President George W. Bush, gained majorities in both the House and the Senate for the first time since 1994. One consequence of this was the passage of the Energy Independence and Security Act (EISA) in 2007. EISA was a key change of course from the existing emphasis on fossil fuel extraction and the former denigration of the power of energy efficiency and conservation. In 2008, the Democrats took even more seats in the House and Senate, and, of course, Barack Obama was elected president.

In the House that year, Henry Waxman, a long-serving congressman from Los Angeles, assumed the chairmanship of the Energy and Commerce Committee. Waxman created a new Subcommittee on Energy and Environment and appointed Ed Markey from Massachusetts its chair. Markey continued to chair the Select Committee on Energy Independence and Global Warming, created in 2006.

Waxman and Markey, working with the full support of the Speaker of the House, Nancy Pelosi, and the new president, quickly commenced hearings on comprehensive energy and climate legislation.

It was not only the political stars in Washington that had seemed to realign for a robust legislative approach to solving the climate crisis. Business and industry were becoming increasingly more vocal and active. The United States Climate Action Partnership (USCAP), for example, was formed at the beginning of 2007. Composed of a number of Fortune 500 businesses based in or operating in the United States, as well as leading environmental organizations, USCAP issued "A Call for Action" in which it stated several principles, among them the need for early action, a recognition of the global dimensions of the problem, the creation of economic opportunity and advantage, and a mandatory and flexible climate program — including "an economy-wide, market-driven approach that includes a cap-and-trade program that places specified limits on GHG emissions."[2] Among USCAP's members are DuPont, Ford, GM, GE, Chrysler, Honeywell, Duke Energy, Shell, Siemens, Weyerhaeuser, and, on the environmental side, the Nature Conservancy, the Natural Resources Defense Council, and the Environmental Defense Fund.

Similarly, a coalition of labor and environmental groups, the BlueGreen Alliance, formed in 2006. It is dedicated to the proposition that a clean-tech economy "has the potential to create millions of jobs, while reducing global warming emissions and moving America toward energy independence."[3] The BlueGreen Alliance is anchored by the United Steelworkers and the Sierra Club, with other members as diverse as the Service Employees International

Union, the National Wildlife Federation, the Laborers' International Union of North America, the Union of Concerned Scientists, the Utility Workers Union of America, and the American Federation of Teachers.

These coalitions and their members supported Waxman-Markey, as did a large number of other business, industry, labor, farm, community, civic, and faith groups. Some highly visible names among these were the AFL-CIO and the UAW; the National Farmers Union and the American Farmland Trust; the U.S. Conference of Mayors and the League of Women Voters; the National Council of Churches, the Catholic Relief Services, the Jewish Council for Public Affairs, and the Evangelical Climate Initiative; as well as American Chemical Society, the American Public Health Association and Physicians for Social Responsibility.[4]

One would think that legislation that had garnered such deep and broad support and was geared, in the words of the bill itself, "to create clean energy jobs, achieve energy independence, reduce global warming pollution and transition to a clean energy economy"[5] would sail through the Senate. Such was manifestly not the case.

A Price on Carbon

At the heart of Waxman-Markey — but certainly not the alpha and the omega — was cap and trade. This approach to reducing greenhouse gas emissions, using a market mechanism to facilitate the required reductions, was also the core of several iterations of draft legislation in the Senate going back to the "Climate Stewardship Act" proposed by Joe Lieberman and John McCain in 2003.

What is cap and trade? It's a "market mechanism" that limits greenhouse gases and allows emitters flexibility in how they achieve the mandated limits. The cap is set by a regulatory authority — for the European Union, it's the European Commission — and requires regulated entities, be they electric power plants, oil refineries, or cement factories, to meet certain prescribed limits to their emissions. The "trade" part of the equation allows an entity to buy or sell credits on an open market. If, for example, an electric utility in Germany achieves reductions below its mandated limit for a set deadline by one hundred thousand tons of carbon dioxide equivalent, it can sell credits for that amount on the market. It will receive whatever price the market is giving. Conversely, should an airline exceed its emission target by seventy-five thousand tons, it can buy credits to make up for that shortfall. Otherwise it will be subject to penalties for the deficit.

In economic terms, greenhouse gases are a "negative externality." The *Economist* defines externality this way: "An economic side-effect. Externalities are costs or benefits arising from an economic activity that affect somebody other than the people engaged in the economic activity and are not reflected fully in prices." Further: "Because these costs and benefits do not form part of the calculations of the people deciding whether to go ahead with the economic activity they are a form of market failure, since the amount of the activity carried out if left to the free market will be an inefficient use of resources."[6]

Lord Nicholas Stern, who led the United Kingdom's pathbreaking review on the economics of global climate change, framed it in these words when the report was launched in October 2006:

> The science tells us that greenhouse gas emissions are an externality; in other words, our emissions affect the lives of others. When people do not pay for the consequences of their actions we have market failure. This is *the greatest market failure the world has seen*. It is an externality that goes beyond those of ordinary congestion or pollution, although many of the same economic principles apply for its analysis. This externality is different in 4 key ways that shape the whole policy story of a rational response. It is: global; long term; involves risks and uncertainties; and potentially involves major and irreversible change.[7]

How do you correct, then, for this unique and massive market failure? According to the *Stern Review*, the Intergovernmental Panel on Climate Change, the United Nations Framework Convention on Climate Change, the European Union, almost all environmental organizations, and the great majority of economists, in order to adequately address the climate crisis, it is critical to set a "price on carbon." There are a number of ways to do this, cap and trade being but one among them.

A tax is another way. There are, of course, enormous political obstacles in the United States to raising taxes, no matter how fair or well constructed they are. This tax aversion is not a phenomenon, certainly, that exists only in

America. At a forum on the question of whether or not a tax on greenhouse gases would work better than cap and trade, the director of the Earth Institute at Columbia University, an internationally renowned development economist, Jeffrey Sachs, said Americans are "neurotic" about taxes. Another commentator, Henry Derwent, head of the International Emissions Trading Association, recounted how he had been briefing a top British official and ticked off a number of the very positive aspects of a carbon tax, but that that worthy high panjandrum said he would never be reelected if he supported such a thing.[8]

In 1993, President Clinton proposed a "Btu tax" to be levied on energy sources based on their heat content. Renewable energy sources such as solar, wind, and geothermal were to be exempted. Although the plan passed in the House in the spring of that year, withering opposition from business, led by the National Association of Manufacturers and the American Petroleum Institute, with a sophisticated and well-financed PR campaign led by Burson-Marsteller, coupled with lukewarm support in the Senate, led Clinton to withdraw the proposal.[9]

Fourteen years later, the then-chairman of the House Energy and Commerce Committee, John Dingell, proposed that "some form of carbon emissions fee or tax (including a gasoline tax) would be the most effective way to curb carbon emissions and make alternatives economically viable."[10] He did not, in the end, pursue a carbon tax vigorously, and his committee subsequently issued several white papers on a cap-and-trade scheme. In 2008, Henry Waxman successfully challenged Dingell and became chair of the committee.

Another reason for a cap-and-trade approach to setting a "price on carbon" — aside from the political need, in Dingell's words, "to overcome ideological Republican opposition to all forms of taxation,"[11] — is the importance of setting a firm cap. A tax does not give the assurance that you will meet the level of emission reductions necessary to effectively mitigate climate change. A tax, in the words of environmental economists, gives you "price certainty," but cap and trade gives you "volume certainty."

Yet another rationale for cap and trade is that the politics across the board can be managed much more easily than with a carbon tax. In fact, one distinguished environmental economist, Robert Stavins, said, just prior to the passage of Waxman-Markey by the House of Representatives, that "the politics of cap-and-trade systems *are* truly quite wonderful, which is why these systems have been used, and used successfully."[12] There was, as would be expected, a tremendous fight for advantage among various industries and political constituencies in the construction of a cap-and-trade vehicle. Stavins writes, however, that "it should not be forgotten that the much-lamented deal-making that took place in the House committee last week for shares of the allowances* for various purposes was a good example of the useful, important, and fundamentally benign mechanism through which a cap-and-trade system provides the means for a political constituency of support and action to be assembled (without reducing the policy's effectiveness or driving up its cost)."[13]

There is another way to put a price on carbon, albeit an implied one: regulation-based policies. Barack Obama, his EPA administrator, Lisa Jackson, and nearly every environmental voice, including Al Gore, made it plain that the best solution to the climate crisis for the United States would be federal legislation and robust American participation in an international agreement through the UN process. Obama and Jackson, however, also made it plain, given the urgency and immediacy of the problem, that they could not and would not rely on Congress to pass legislation. The administration moved aggressively on a number of fronts, including creating a regulatory framework, under the Clean Air Act, to address greenhouse gases. We will see how this framework has been constructed in the next chapter.

Cap and Trade: Death in the Senate

With the House having done its work, the emphasis shifted to the Senate. There were, as we have noted, scores of heavy-hitting organizations supporting cap and trade. There was even bipartisan precedent in the Senate, cap-and-trade bills having been offered by conservative Republicans like John McCain in 2003, and, in 2007, by John Warner, who put his name to the Lieberman-Warner Climate Security Act.

The Democratic Party gained a majority in the Senate in 2006. In so doing, it installed an outspoken progressive supporter of action on climate

---

* An allowance is the designated amount of pollutants that a source may emit under a cap-and-trade system, measured in tons of CO_2eq.

change, Barbara Boxer, as chair of the Environment and Public Works Committee. Jeff Bingaman, a lower-key but committed proponent of sustainability, became chair of the Energy and Natural Resources Committee. However, even though there was some bipartisan support for climate and energy action, there was what turned out to be a fatal combination of general Republican opposition to anything hinting of tax increases—the *Wall Street Journal* deriding Waxman-Markey as "The Cap and Tax Fiction"[14]—and entrenched resistance by both Republican and Democratic senators whose states relied on the coal or oil industry or where the preponderance of electricity was generated by coal.

What is particularly ironic about Republican opposition to cap and trade is that it is manifestly a market mechanism that allows industry to avoid a strict regulatory regime that market fundamentalists so thoroughly loathe. In fact, cap and trade was given birth by President George H. W. Bush and Congress in 1990 as a highly successful, cost-effective scheme that radically reduced America's acid rain problem. In 1988, Environmental Defense Fund president Fred Krupp contacted Bush's counsel, Boyden Gray, and proposed the scheme to deal with the emissions of sulfur dioxide and nitrogen oxides from electrical and industrial power plants. EDF was, among environmental organizations, an early convert to the idea of working *with* business and industry to get at various pollution problems. President Bush embraced the idea, making the proposed emission reductions even larger than environmentalists had been proposing.[15]

Thus, cap and trade was at the heart of the acid rain title in the 1990 Clean Air Act Reauthorization. It became a success far beyond even the expectations of the environmental economists who had worked with the White House and Congress to structure it, with emissions dropping more quickly and less expensively than anticipated. Figure 5.1 is a chart detailing the speed and cost-effectiveness of the federal acid rain program.

Europe, seeing the success of the American acid rain program, adopted cap and trade as its primary market mechanism for reducing greenhouse gases. Senator Warner, among many others, noted the European example. "We've had experience that this cap-and-trade system has worked successfully. It is working in the European Union countries, so there [is] some precedent for doing it."[16]

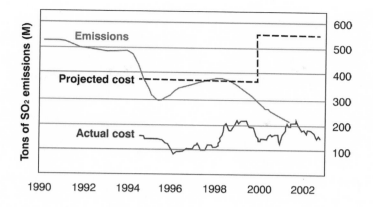

5.1 The acid rain experience: unprecedented environmental protection at unmatched cost efficiency. "The Cap and Trade Success Story," Environmental Defense Fund

Senator Boxer held numerous hearings on climate change from 2007 when she assumed the chair of the Environment and Public Works Committee. Her hearings evidenced high measures of depth and breadth of support for cap and trade and the complementary measures necessary to achieve lasting reductions in greenhouse gases and setting the United States on a permanent course for growth in green tech and jobs. In the fall of 2009, Boxer and former presidential candidate John Kerry introduced the "Clean Energy Jobs and American Power Act." It was analogous in many respects to Waxman-Markey, giving environmentalists some hope that a final vehicle could be negotiated between the House, Senate, and White House, and that President Obama could sign significant legislation into law in 2010.

But the political climate itself had become much too overheated for any sort of simple and easy ride. When Boxer's committee proceeded to "mark up" the draft, the entire Republican membership of the committee boycotted the working session, an ill omen indeed. The bill passed out of committee without any Republican participation in the vote, but negotiations were soon under way for another vehicle that would, hopefully, bring enough votes on board to allow legislation to move forward. Kerry, working with former vice-presidential candidate Joe Lieberman, now an Independent (but caucusing with the Democrats), and with Republican Lindsey Graham, were crafting a draft that would come to be known as the American Power Act.

By May of 2010, the details of the draft were released, but by the end of

July, the bill was, for all intents and purposes, dead. Senate Majority Leader Harry Reid announced that the legislation would have to wait at least until the fall for action. Given that it was an election year for one-third of the senators and all of the representatives, it was not felt likely that legislation could be passed. To add insult to injury, the prospects for considerable Republican gains in the fall elections might well doom significant federal climate legislation for the entirety of the 112th Congress, from 2011 through 2012. This turned out to be the reality.

Some environmentalists blamed the president. One of the most prominent observers in Washington, and an outspoken and eloquent advocate for strong, timely, and smart action, Joe Romm, went so far as to say that the president had shown a "stunning indifference to the defining issue of our time" and that if Obama was not able to effect climate and energy legislation on his watch, he would have had a "failed presidency."[17]

New York Times columnist and Nobel laureate in economics Paul Krugman accused those senators who are beholden to coal or oil money and the ones who are afraid to buck the right-wing orientation of their party. "Greed, aided by cowardice, has triumphed. And the whole world will pay the price."[18] Ross Douthat, another *New York Times* columnist, noted in his obituary for the legislation that "cap-and-trade's backers are correct to point the finger rightward. If their bill is dead, it was the American conservative movement that ultimately killed it."[19]

There is another salient reason why this legislation died. One prominent political analyst put it this way: "But one reason towers above all others — the dysfunctionality of the Senate."[20] Aside from the arcane rules, very much including the filibuster,* or even the threat of a filibuster, which allows a minority to block voting on any measure, as well as the Senate's super-saturation by special-interest money, it must be remembered that the Senate itself is an inherently undemocratic institution by virtue — or lack thereof — of the fact that it violates one of the foundational principles of democracy, namely "one person, one vote." Rhode Island and Iowa are as equally represented in the

---

* According to the U.S. Senate, filibuster is "an informal term for any attempt to block or delay Senate action on a bill or other matter by debating it at length, by offering numerous procedural motions, or by any other delaying or obstructive actions." The term itself derives from a Spanish word for pirate, an apt etymology.

Senate as California and Texas, Wyoming and South Carolina as New York and Illinois. This is, according to one distinguished political scientist, "a profound violation of the democratic idea of political equality among all citizens."[21] Thus, even a small minority of senators representing a relatively small proportion of the American populace can block whatever legislation it opposes. That is, for instance, how the South blocked any advance on civil rights in the United States for decades, even anti-lynching legislation.

The concatenation of Republican anti-environmentalism and fear (and no doubt loathing), plus intransigence from Democratic senators from states where coal and oil are king, brought climate and energy legislation crashing down, like the *Hindenburg*, in flames.

Al Gore named several reasons for the failure, among them that "the influence of special interests is now at an extremely unhealthy level. And it's to the point where it's virtually impossible for participants in the current political system to enact any significant change without first seeking and gaining permission from the largest commercial interests who are most affected by the proposed change."[22]

Indeed, even as the grenades and Molotov cocktails were being lobbed to block the enactment of effective climate and energy legislation, a number of attempts were being made to curtail or eliminate EPA's existing authority under the Clean Air Act to regulate greenhouse gases. In June of 2010, four Democrats joined all the Republicans in the Senate in voting to block the EPA. The measure failed, but Jay Rockefeller, Democrat of West Virginia, brought the measure before the Senate in the 112th Congress, as did the new chair of the House Energy and Commerce Committee, Fred Upton. So far, these attempts have failed, and President Obama vowed to block any proposal that might surface from the Congress.

At the end of the day, though, a bad, watered-down Senate bill was going to be worse than no bill. Waxman-Markey, decried by the Right and, in many cases, by the Left, was essentially sound in formulating not only a politically and economically astute approach to cap and trade, but also in creating strong sections on renewable energy and energy efficiency, the smart grid and transmission, transportation, research and development, green jobs, and agriculture and forestry. However, the American Power Act was in many ways, before it crashed and burned, a giveaway to the oil and coal interests, the nuclear power industry, and even stripped the EPA of its statutorily mandated regulatory authority under the Clean Air Act *and* the states of their right to act on their own initiatives to address energy and climate concerns.

Finance Takes an Interest

Many people, in the United States and internationally, regard the failure of Congress to pass climate legislation as a fatal flaw in the fabric of our efforts. Many environmentalists also think that without a specific American federal law that mandates a reduction in greenhouse gases, we will never be able to avoid the looming catastrophe. It is certainly entirely true that if we do not perform a number of important actions vigorously, efficiently, and soon, we are going to be experiencing what ecologists, public health experts, economists, and human security analysts, among others, describe as global change on a scale that is unprecedented, certainly during the time of humanity's tenure here over the past several hundred thousand years.

But as we have seen, there are extraordinary breakthroughs taking place in clean tech, as well as in policy. Some very important advances have been taking place in the business and financial communities as well. Mark Twain wrote of one character that "she was wise, subtle, and knew more than one way to skin a cat."[23] In other words, if some political systems won't participate in generating an appropriate response to the looming climate catastrophe, then other institutions and entities will.

The business world expected an American carbon market mandated by an economy-wide cap-and-trade law. It did not, however, hold back on developing other markets while it waited. The Kyoto Protocol created "flexible mechanisms" for participating developed economies to use. The Europeans, for one, have moved smartly in creating their Emissions Trading System (EU ETS) and getting it up and running.

Recognizing cap and trade as an efficient and equitable vehicle for greenhouse gas reductions, and inspired by the success of the U.S. acid rain program, the EU launched its program in 2005. The EU program regulates carbon dioxide and nitrous oxide emissions, accounting for 40 percent of the EU's GHGs, with over eleven thousand facilities such as power plants, oil refineries, and heavy manufacturing facilities making steel, cement, and ceramics, and other products covered. Aviation was brought into the scheme in 2012. More industries and more greenhouse gases are coming into the system over the next few years. One of the flaws in the early days was that more allowances were allotted to industry than were necessary, depressing the prices on the markets. As the system has advanced and matured, though, adjustments have been made. The adjustments are ongoing, as is the tightening of the caps necessary to achieve the target of a 20 percent reduction of greenhouse gases from 1990 levels by 2020 that the EU has set for itself.[24]

The EU ETS represents the lion's share of the global carbon market. The global carbon market saw seven billion tons of allowances and credits for carbon dioxide equivalent (CO_2eq) traded in 2010 with a value of €92 billion (US $123 billion). Of this, the EU ETS accounted for 5.2 billion tons, worth €72 billion ($96 billion). The volume of the global market increased by a factor of nine from 2005, when the first phase of the EU ETS began, through 2010.[25]

American Players

The EU ETS has been the principal actor in the still very young carbon markets, but there are other players on the scene. There is, for instance, the Regional Greenhouse Gas Initiative (RGGI) in the United States. RGGI — pronounced *reggie* — is a compact of ten eastern states, from Maryland to Maine, that have instituted cap and trade for electric power plants. They allocate the allowances — the "permits" to generate a capped volume of carbon dioxide — through an auction process. The auctions generate revenue for the participating states. The first auction took place on September 25, 2008. Through September of 2011, proceeds from the auctions totaled more than $900 million. Over 80 percent of that money has gone to underwriting clean energy programs such as supporting energy efficiency retrofits and accelerating renewable energy projects.[26]

Beyond the capping of emissions and the generating of revenue for clean energy, one of the principal purposes of RGGI is to serve as a proving ground for other regional programs and, eventually, it is hoped, a U.S. arrangement.

Other regional initiatives are just now gathering steam. One is the Midwestern Greenhouse Gas Reduction Accord. This regional compact has six states and one Canadian province as members and four observer states. In

November 2007, the governors of the member states and the province entered into an agreement to explore the reduction of greenhouse gases through a regional cap-and-trade system. In May 2010, an advisory panel made its recommendations on how to proceed. The regime would cover emissions of the six Kyoto gases (carbon dioxide, methane, nitrous oxide, hydrofluorocarbons, perfluorocarbons, and sulfur hexafluoride) and apply to emissions from electrical and industrial power plants, manufacturers, and fuels for buildings and surface transportation. A final agreement, though, has not yet been reached.[27]

California's program, however, is well advanced. Governor Arnold Schwarzenegger signed a law in 2006, AB 32, creating a cap-and-trade program for the state. In 2010, an attempt was made to roll back the law via a referendum. The battle over AB 32 became a focal point for environmentalists, participants in the carbon markets, and clean-tech entrepreneurs and investors all over the world. The voters soundly rejected the attempt to stop California's key climate-change-fighting vehicle, 61 percent to 39 percent, with many liberal and moderate Republicans among those voting to keep AB 32 in place. A month later, the California Air Resources Board (ARB) endorsed the cap-and-trade program. ARB chairman Mary D. Nichols said, "This program is the capstone of our climate policy, and will accelerate California's progress toward a clean energy economy."[28] Surviving a court challenge, in October 2011 the ARB adopted the final regulations to put cap and trade in place.[29]

The program will cover 360 kinds of businesses, with the initial phase beginning in 2012 with all major industrial sources, including utilities, under regulation, and then a second phase that starts in 2015 covering transportation fuels, natural gas, and other fuels. The aim is to return GHG emissions to 1990 levels by 2020.

The importance of California taking this initiative cannot be overstated. It is the most-populous U.S. state and the ninth-largest economy in the world, with a GDP of $1.9 trillion in 2010, ranked just below Brazil and above Canada. California has headquarters for fifty-seven of the Fortune 500 companies. The Golden State is a global nexus for clean tech, attracting, for instance, $11.6 billion in clean-tech venture capital from 2006 through 2010, 24 percent of the world total.[30] It also tops one list for U.S. clean energy leadership, as measured by technology, policy, and financing trends and best practices.[31] For California to stay on course and expand its drive toward a low-carbon economy is critical to the rest of the world's efforts.

California also anchors the Western Climate Initiative (WCI), an agreement among the Golden State and several Canadian provinces to work cooperatively to reduce greenhouse gases. Formed in 2007, the WCI originally had seven participating states and four provinces, but six states withdrew in late 2011 to participate in a new initiative, North America 2050. California, Quebec, Ontario, Manitoba, and British Columbia remain and have established a nonprofit corporation to support their efforts. A comprehensive GHG reporting system is under way for the participants, and the first compliance period began in 2012 and runs for three years. The WCI covers emissions from electricity, industry, transportation, and residential and commercial fuel use. Overall, the program is targeting a 15 percent reduction in GHGs from 2005 levels by 2020. The four participating Canadian provinces represent nearly three-quarters of the Canadian economy.

The WCI members are making an enormous commitment to clean energy and to a considerable reduction in greenhouse gases. The politics don't work for the U.S. government as yet, but these subnational governments, with the support of their publics, as evidenced by the convincing results of the referendum in California in November 2010, are moving forward. There appears to be a recognition among these populations both of the risks involved in rising temperatures and that clean tech offers enormous potential for jobs and investment, money and energy savings, and economic growth.

The International Scene: Beyond Europe and the United States

Felipe Calderón, president of Mexico, said it quite succinctly prior to the opening of COP 16 in Cancún: "Fact: The only sustainable path to growth is a low-carbon path."[32]

New Zealand's Climate Change Response Act of 2002 set up its emissions trading scheme. It is a truly economy-wide system, covering everything from fishing and forestry to transportation fuels and energy production. In announcing a review of the ETS in order to fine-tune it, the climate change issues minister, Nick Smith, said, "All the international evidence confirms that pricing emissions is the most efficient way of addressing climate change."[33]

A thousand miles across the Tasman Sea in Australia, politics have been greatly influenced by climate change in recent years. A Labor government came to power in 2007 partly because Australians had become tired of Conservative prime minister John Howard's foot-dragging on climate legislation. Australia, after several years of political and parliamentary maneuvering, finally passed comprehensive climate and energy legislation out of its House of Representatives in the fall of 2011.[34] With this final major hurdle successfully passed, a full package of measures, including a carbon tax, went into effect in 2012. The carbon tax will become a cap-and-trade system in 2015, and Australia will then enter into the international carbon markets. This will give a boost to the nascent international market and allow the Australians more flexibility in meeting their obligations.[35] The electric utilities, perhaps ironically, were among the most affected by the delay in creating a market-based framework. They complained that they couldn't plan for the future until they knew the framework in which they would be working.[36] Prior to the passage of the legislation, though, Australia had instituted a Renewable Energy Target requiring that 20 percent of electricity come from renewables by 2020.[37] Australia also is promoting "A New Car Plan for a Greener Future" in which billions will be spent in public money "to help the automotive industry to prepare for a low carbon future and to make the industry indispensable to global markets and supply chains."[38]

South Korea and Japan have proposed cap and trade, but powerful industry interests are holding things up. If programs started there in 2013, the Japanese carbon market could be worth €106 billion ($141 billion) in 2020, while South Korea's could reach €56 billion ($75 billion).[39] While the Japanese, with the world's third-largest economy, have put cap and trade on hold, they are instituting higher taxes on coal and oil and are strengthening their Renewable Portfolio Standard (RPS), like Australia's Renewable Energy Target, which mandates a certain amount of electricity be generated by renewables.[40] South Korea is creating its own RPS, set to begin in 2012 and reach 10 percent by 2022.[41]

Tokyo, however, has not waited for the national government to act. On April 1, 2010, it commenced its own cap-and-trade program. Tokyo's goal is to achieve 20 percent emission reductions below 2000 levels by 2020. The program is different from others because it seeks to promote green energy use by consumers "downstream" rather than require it from energy producers "upstream." In other words, the program requires about fourteen hundred commercial buildings, factories, and other large users of power—accounting for 1 percent of the country's emissions—to cut their output of greenhouse gases by about 6 to 8 percent during the first compliance phase, then 17 percent later on, from 2015 to 2020. Improving building energy efficiency will be the principal modus operandi for achieving these reductions.[42]

China is the largest emitter of greenhouse gases in the world. For carbon dioxide alone, China released 7.5 billion tons in 2009—24 percent of the global total.[43] China has, however, made a commitment through the Copenhagen Accord and the Cancún Agreements, not to mention as a matter of domestic policy, to reduce its "carbon intensity"* from 2005 levels by 40–45 percent by 2020. In order to do this, China will impose binding emissions limits. This will require jurisdictions and industry facilities to accurately measure their carbon dioxide output. As one expert notes, it is important to remember "the significance of the carbon intensity target for creating the proper framework and incentives for reducing emissions."[44] While reducing carbon intensity is certainly not the be-all and end-all, given the continuing white-hot pace of economic expansion there, China has been accelerating its program of improvements on energy efficiency, build-out of renewable energy and smart grid infrastructure, and urban mass transit and high-speed rail.

China's leading climate change official, Su Wei, has indicated that a domestic carbon trading platform is under study.[45] Meanwhile, voluntary environmental exchanges were established in Beijing, Tianjin, and Shanghai in 2008. These exchanges are serving as pilot programs in China for emission trading. One prominent environmental economist, Lord Stern, has predicted that China will have cap and trade by 2015.[46] The central government, in fact, ordered two provinces and five major cities in early 2012 to set caps in anticipation of local carbon markets.[47]

Brazil has ambitious goals for reducing its greenhouse gas footprint, mostly centered on initiatives to curb deforestation. Reducing emissions from deforestation and forest degradation (REDD) is a critically important set of initiatives for countries like Brazil and Indonesia. But Brazil's rapidly grow-

---

* Carbon intensity is the energy-related carbon dioxide emissions per unit of economic output.

ing economy was responsible for putting out 374 million tons of CO_2 in 2007, slightly more than Saudi Arabia and less than France.[48] As in China, policy makers are considering introducing a domestic cap-and-trade system.[49]

India has two very young programs that trade in renewable energy and energy efficiency certificates. Mexico has a new program for voluntary GHG accounting and reporting that will cover 80 percent of emissions by 2012. The program is in place to build capacity for future participation in the carbon markets.[50]

With China and other leading developing economies like Brazil, India, and Mexico on board, there will be a much bigger pie for the carbon markets.

Prognosis

Overall, the outlook for cap and trade is hopeful. One analyst said that, for the United States, carbon markets could be worth $2 trillion in transaction value within five years of starting trading. "That would make it the largest physically traded commodity in the U.S., surpassing even oil."[51]

And no less astute an observer of international finance than George Soros thinks the idea of different carbon markets arising is better than one UN-mandated, global system:

> Even as the top-down approach to tackling climate change is breaking down, a new bottom-up approach is emerging. It holds out better prospects for success than the cumbersome United Nations negotiations. Instead of a single price for carbon, this bottom-up approach is likely to produce a multiplicity of prices for carbon emissions. This is more appropriate to the task of reducing carbon emissions than a single price, because there is a multiplicity of sectors and methods, each of which produces a different cost curve.[52]

Mindy Lubber, president of Ceres, a network of investors, environmental organizations, and other public interest groups that works with companies and investors to address challenges like climate change, is bullish on the future of emissions trading because of what she perceives as a growing chorus from business for a market-based framework in which to address reductions. "I actually believe that over the next few years that some of the loudest spokespersons and proponents for putting a price on carbon and a cap on carbon — and it will look different from the comprehensive cap-and-trade bill that we lost, monumentally, in the last Congress — will be business, financial, and public health leaders who are saying that for the future of our public as well as our economy we need to be acting on climate change."[53]

The Clean Development Mechanism

One of the key "flexible mechanisms" to come from the Kyoto Protocol is the Clean Development Mechanism (CDM). The CDM allows entities that are involved in programs such as the European Union Emissions Trading System, as part of a cap-and-trade regime, to buy "offsets" for their emissions. Given the principle that anthropogenically induced greenhouse gases are everywhere equally responsible for climate forcing, then programs that reduce their emission, wherever they may be under way, will be equally useful in mitigating the problem. In other words, the CO_2 from a coal-fired power plant in Italy or the methane leaking from a landfill in Japan each have a calculable impact on the climate system; and because the system is equally affected wherever the source of the greenhouse gas is located, the total CO_2eq emitted, no matter the location of the source, can be reduced or offset.

You can quantify and benefit from reductions wherever they may be occurring, so offsets can come from anywhere. The CDM allows developing countries that have greenhouse gas mitigation initiatives, like Bus Rapid Transit in Bogotá or a brewery in South Africa that has switched fuels from coal and oil to natural gas, to monetize the reductions inherent in these programs by being granted Certified Emission Reduction (CER) credits, equivalent to a ton of carbon dioxide. The CERs can be sold directly or on the carbon markets for use by developed country entities that need to meet their reduction targets.

There is, of course, a complex process for registering kinds of projects and specific initiatives. This is administered by the UN Framework Convention on Climate Change's CDM Executive Board, which itself has a number of panels, including those for methodologies, accreditation, and registration and issuance.

World Bank analysts now expect nearly one billion CERs to have been issued through the end of 2012, the end of the Kyoto commitment period.[54]

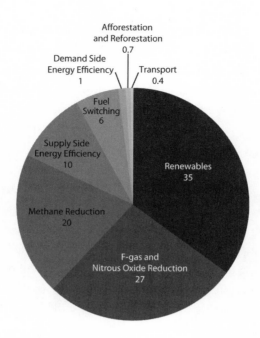

5.2 CERs expected until 2012 from CDM projects in each sector. Graphic by Lorraine Simonello, based on data from the Risoe Centre on Energy, Climate and Sustainable Development of the United Nations Environment Program

This is, unfortunately, a figure significantly lower than the 2.7 billion that the UNFCC itself projects will have been issued.[55] However, looking further into the future, to 2020, one projection is that 3.6 billion CERs will have been issued by that time.[56]

The pie chart in figure 5.2 shows CERs expected to 2012 from CDM projects in each sector.

By the beginning of 2011, actual CER production reached half a billion, issued to 855 projects. One of the critical underlying raisons d'être for the CDM has been to funnel investment to sustainable development projects. Thirty-seven developing countries had been given CERs by the beginning of 2011. One of the problems with the CDM, however, is that only a few countries have been beneficiaries of the lion's share of the CERs issued. Of those issued, 92 percent have gone to China (53.8 percent), India (16.5 percent), South Korea (12.25 percent), and Brazil (9.3 percent).

It is very much on the agenda for the parties to the framework convention to expand, deepen, and strengthen the CDM program as a critical adjunct to sustainable development. We will look more closely in Chapter 7 at the CDM, how it boosts sustainability, and the many "co-benefits" these types of projects bring in terms of lower energy costs, pollution reduction, jobs, and other factors.

Joint Implementation

The third in the Kyoto Protocol's portfolio of flexible mechanisms, after emissions trading and the Clean Development Mechanism, is Joint Implementation (JI). Where the CDM is meant to channel investment from the developed world nations to the developing, JI is structured to allow a shared effort between two or more Annex I nations. These include the highly developed economies and the so-called "economies in transition" (EIT), the former Soviet states in Europe and Soviet bloc countries. The EITs have made commitments under the Kyoto Protocol, but because the devastating economic downturn that accompanied the fall of the Soviet Union generated a concomitant reduction in greenhouse gas emissions, there are surpluses available to them for sale. The global recession that began in 2008 exacerbated the situation in these countries, providing further surpluses.

Each Annex I country is allocated Assigned Amount Units (AAU), the amount of GHG, corresponding to a ton of CO_2eq, they can emit according to their Kyoto commitment. The former Soviet-bloc EITs are beginning to use their surplus AAUs for sale or as collateral for investments by other parties in projects in their countries. Ukraine, for instance, is using its AAU reserves as collateral in a JI scheme that will cut emissions of methane by upgrading Ukraine's gas distribution networks. The project developers have agreed to perform the repairs with the proviso that twenty million AAUs be provided as collateral.[57]

JI projects earn Emission Reduction Units (ERUs), just as CDM projects earn CERs. There were 375 projects in the pipeline through 2012. Russia, for example, working with Switzerland, has a project that will destroy hydrofluorocarbons and sulfur hexafluoride, two industrial gases regulated under Kyoto. This initiative is expected to earn five million ERUs. Another project is Bulgaria and Denmark working together to build a small hydropower station that should generate 85,000 ERUs.[58] Nearly a quarter of these projects are for renewable power, but they will produce less than 10 percent of the ERUs overall.

The JI is a smaller program than the CDM, certainly, but it is nevertheless generating momentum on investment in reducing greenhouse gases while increasing renewable energy and energy efficiency in the host countries, among other worthwhile ends.

The Voluntary Markets

Beyond the "compliance" markets (EU ETS, RGGI, etc.) are the voluntary markets where companies and individuals can find offsets for their emissions. With pilot trading programs in places like China and Mexico, international exchanges like the Chicago Climate Exchange, and with over-the-counter trading, and companies seeking to ease their way into the game as players, the voluntary carbon markets have been a testing ground and an incubator. They have also been an important venue for proving critical means for the measurement, reporting, and verification (MRV) of greenhouse gas mitigation projects.

There is thriving activity globally in project development in the form of forest protection, renewable energy, energy efficiency, and methane destruction, among other sectors. Although the total market value decreased as a consequence of the global recession, from $728 million in 2008 to $387 in 2009 (on volume of 127 million tons of CO_2eq versus 94 million tons CO_2eq),[59] the outlook for the future remains positive. As more companies become involved in "pre-compliance" activities, in anticipation of their full participation in regulated environments like the Western Climate Initiative and others, and as international programs to counterbalance the many pernicious impacts of deforestation and forest degradation advance, the voluntary markets are expected to grow in importance. They will not approach the compliance markets in volume or value, but they are crucial to bringing more participants to the carbon markets and expertise to the field.

Economic Consequences of Climate Change

There are, as we have seen, opportunities for the financial industry in the carbon markets and, as we will see, many more ways that businesses can prosper and job growth can be created in the clean-tech economy. However, there are extraordinary risks to businesses and to our societies generally, given the many impacts that climate change has already caused and will continue to cause. These risks are not going to diminish anytime soon.

In 1990, the Intergovernmental Panel on Climate Change issued its First Assessment Report. The responsibility of Working Group II was to outline "the environmental and socioeconomic implications of possible climate changes over the next decades caused by increasing concentrations of greenhouse gases."[60] The panel identified a number of impacts with economic consequences. "The socioeconomic consequences . . . will be significant, especially for those regions of the globe where societies and related economies are dependent on natural terrestrial ecosystems for their welfare."[61]

More and deeper economic analyses were made by the IPCC in its three subsequent assessment reports. The alarm had clearly been sounded, but it was not until the fall of 2006 that the world really heard the message of the dire economic risks inherent in climate change. It was then that the *Stern Review on the Economics of Climate Change* was launched by the government of the United Kingdom. In the summer of 2005, Gordon Brown, then chancellor of the exchequer, commissioned Sir Nicholas Stern to conduct a broad review of the economic implications of climate change, including looking at both the risks and the costs to mitigate the causes and adapt to the impacts. Stern had been chief economist to the European Bank for Reconstruction and Development and then to the World Bank, and a top British treasury official.

The seven-hundred-page *Stern Review* considered several scenarios, based on scientific findings, and found that world GDP could be diminished by from 5 to 20 percent annually. "Our actions over the coming few decades could create risks of major disruption to economic and social activity, later in this century and in the next, on a scale similar to those associated with the great wars and the economic depression of the first half of the 20th century. And it will be difficult or impossible to reverse these changes."[62] That is stark.

Stern and his colleagues also made it perfectly clear that the costs involved in reducing greenhouse gases and transitioning to a cleaner, cooler energy economy were not only manageable—they were an excellent investment.

The *Stern Review* was a watershed. Although the IPCC had addressed economic risks, as had others, the UK report lent gravitas and directed a spotlight to the issue. The highly articulate Sir Nicholas, now Lord Stern, and his team gave numerous presentations and interviews to promote the review's findings. The media coverage in the fall of 2006, following the release of Al Gore's *An Inconvenient Truth* in the spring, was intense. (See figure 1.2, on English-language newspaper coverage, in Chapter 1.)

In bestowing its Frontiers of Knowledge Award for 2010 to Lord Stern, the BBVA Foundation, endowed by one of the world's largest banks, noted that the idea that the economic impacts of climate change would vastly outstrip the costs required to meet the challenge early and directly had "fundamentally changed the international climate change debate and stimulated action."[63]

Investors and Businesses See the Risks

Several months after the *Stern Review*'s launch, John Llewellyn, a longtime top economist with the Organisation for Economic Co-operation and Development (OECD) and then an economist with Lehman Brothers, released another seminal report. Llewellyn noted that "in the world of business and finance, climate change has developed from being a fringe concern, focusing on the company's brand and its Corporate and Social Responsibility, to an increasingly central topic for strategic deliberation and decision-making by executives and investors around the globe.[64] He further said, "Businesses are likely to be affected both by climate change itself and by policies to address it through: regulatory exposure; physical exposure; competitive exposure; and reputational—including litigational—exposure."[65]

Nonprofit organizations like the Carbon Disclosure Project (CDP) play an important role in bringing a focus to the risks that corporations are incurring from climate change. The first step in managing these risks is measuring the greenhouse gas output of a business's operations, "because in business what gets measured gets managed."[66] CDP gathers this data from three thousand companies worldwide and promulgates the information to governments, investors, and other businesses, academics and think tanks, and NGOS. The Climate Registry is another nonprofit doing similar work with a focus on North America. In order to create a global standard for this sort of reporting, the Climate Disclosure Standards Board (CDSB) was formed in 2007 at the World Economic Forum. Two of the members of the founding board said, "Climate change and the implications on business process and disclosure are finally becoming the topic of discussion that they deserve to be." Mindy Lubber, another founder of the CDSB, said at the time: "This initiative is a key step to improving and standardizing company disclosure on the risks and opportunities from climate change, whether from new regulations, physical impacts or growing global demand for clean technology products."[67]

Lubber is the head of Ceres. Ceres, like the U.S. Climate Action Partnership and so many other highly effective organizations dedicated to sustainability, is a coalition of scores of groups from the public interest sector as well as investment institutions and foundations. Ceres works on a number of fronts, but one of its principal jobs is making sure that corporations know that investors are watching and concerned. It works to promote awareness in the investment community of the risks and the opportunities presented by climate change. The ninety institutional investors and financial firms of the Investor Network on Climate Risk (INCR), launched by Ceres, have some juice: the members collectively manage nearly $10 trillion in assets. When INCR talks, Wall Street and corporate boards listen.

But are Wall Street and industrial and commercial enterprises getting the message? Echoing John Llewellyn, Lubber notes: "There is no question in my mind that we are no longer debating the issue of whether or not climate change is real and has a real financial impact within the corporate community." For example, the interest in and attendance at climate and sustainability sessions at the World Economic Forum have continued to mushroom over each of the past several years. "It is seen as a world class environmental, national security, public health and financial issue within the financial and corporate sectors. It is radically different from five years ago."[68]

One way to pressure a company is to bring shareholder proposals calling for greater attention to climate risk to a vote at annual general meetings. As the number and strength of these proposals have grown each year, many more of them are being addressed by the companies by their agreeing, without the actual vote, to meet the terms of the proposals.

Other groups like Ceres keep the focus on climate change with public statements, shareholder activism, conferences, and by engaging directly with companies. The Institutional Investors Group on Climate Change (IIGCC) has over sixty-five members in Europe managing around €6 trillion ($8 trillion). In November of 2010, Ceres, IIGCC, and others issued a call to world governments: "Take action now in the fight against global warming or risk economic disruptions far more severe than the recent financial crisis." The statement was signed by 259 investors from the United States, Europe, Asia, and Australia with collective assets totaling over $15 trillion.[69]

Even with all the focus and energy that is being brought to bear, the news is not all positive. A survey of top investment managers by Ceres revealed that they are not adequately seeing and addressing the risks, many going so far as to say they don't believe that climate change is financially "material" to their investment decisions.[70]

Al Gore refers to fossil fuel and related inventory and investments as "subprime carbon assets." With the advent and advance of the movement toward putting a price on carbon, Gore thinks "the owners of these assets will soon face a reckoning in the marketplace. They are in roughly the same position as the holders of subprime mortgages before they realized the awful mistake they had made."[71]

The Insurance Industry and the Banks

The insurance industry, perhaps more than any other, is disposed and fully competent to understand and assess the risks in climate change. Hurricanes, sea-level rise, higher-than-usual rainfall events, as well as droughts and catastrophic heat waves, and even devastating fires in forests and farmlands, are all among natural catastrophes driven by or exacerbated by climate change. One of the world's largest insurance industry players, Munich Re, in a report on a very difficult year, 2010, said: "The high number of weather-related natural catastrophes and record temperatures both globally and in different regions of the world provide further indications of advancing climate change."[72] Lord Peter Levene, chairman of Lloyd's, noted that "we cannot risk being in denial on catastrophe trends. We urgently need a radical rethink of public policy, and to build the facts into future planning."[73]

The insurance industry has been seeing to its own house. ClimateWise, for instance, is an initiative to perform advanced risk analysis, support good public policy decisions, and promote awareness of the issue in the industry, among other things. Founded in 2006, it now has forty members around the world, some of the biggest names in insurance, like AXA, Allianz, Aviva, Swiss Re, and Lloyd's.[74] One of its members, Zurich Financial, issued a report fully acknowledging the risks but asserting that "the insurance industry can play a central role in dealing with climate change, and to articulating a framework under which public policy should be framed in order to avoid unsustainable risk creation and accrual in our approach to climate change."[75]

A characteristically thorough and lucid report from Ceres surveyed the state of the industry and reported on the huge growth and innovation in products and services that helped businesses reduce not only their exposure to climate risk but their emissions as well. The insurance industry has also been investing in clean tech. Still, although this activity has been vigorous, much more is possible, desirable — and necessary. "The activities described in this report indicate the vast potential for insurers to introduce new climate-friendly products and services through their core business, and to participate in the coming 'green revolution' in the financial markets through their investments and asset management. The challenge will be to ensure that these products are brought to scale in time to have a material impact on what is likely to be the biggest challenge facing the industry in its history."[76]

Meanwhile, six of the biggest banks[77] in the United States (and the world), are signatories to the Carbon Principles. These principles seek to both "reduce the regulatory and financial risks associated with greenhouse gas emissions" and to encourage the power industry in the United States to reduce emissions through, among other things, enhancing energy efficiency, including promoting "regulatory and legislative changes that increase efficiency in electricity consumption including the removal of barriers to investment in cost-effective demand reduction." These banks also explicitly endorse a market-based approach to regulating carbon.

Working with some leading environmental organizations and several of the biggest investor-owned utilities, these banks apply an "enhanced diligence process" to financings for coal-fired power plants.[78] As we will see in the next chapter, coal's role as the fuel for new power plants in the United States has been radically curtailed in recent years. Part of the reason for this pullback on coal is the influence of the banks.

Reducing Risk

As the insurance industry has shown, the next step after measuring and reporting greenhouse gases and climate risks is to get at the task of reducing emissions and dealing with the risks. Corporations have been doing this in increasingly more assiduous and innovative ways. They are certainly being helped along by nonprofits, consulting firms (large and small), international trade organizations, and, in many cases, government agencies — multilateral, national, and local.

There are scores of these initiatives. Here are some examples:

· The Climate Group is a London-based international NGO that works with its corporate and government members on developing technologies and policies. Members are involved in everything from aviation to IT, retail, and energy. Governments on board range from the cities of New York and London to states and provinces from Australia, Europe, South and North America.
· The World Business Council for Sustainable Development has member companies in Africa, Asia, Europe, Latin America, North America, and Oceania. Its sustainability initiatives include those on water, energy efficiency in buildings, sustainable forest products, cement, electricity, tires, mining, and mobility.
· The UK's Carbon Trust helps companies lower their GHG footprint with expert advice, finance, and accreditation.
· The Clinton Climate Initiative (CCI) works with governments and businesses around the world to maximize energy efficiencies in the urban environment, promote clean energy, and stop deforestation.
· In the United States, the Natural Resources Defense Council (NRDC) and the Environmental Defense Fund (EDF) have both been working for years with businesses to maximize their clean-tech presence

and minimize their environmental impact. Both are venerable, internationally respected environmental organizations. NRDC's Center for Market Innovation, for instance, promotes investments and new financing mechanisms. EDF works on the policy side with business, being, for example, one of the architects of the U.S. Climate Action Partnership, but also helps effect meaningful breakthroughs on the ground. EDF's work with Walmart, the world's largest retailer, on supply chain sustainability, renewable energy, and alternative fuels for its vehicles has been a critical initiative. (EDF's role in eliminating most of the eleven proposed coal-fired power plants slated for construction in Texas as part of the deal for the acquisition of a Texas utility will be outlined in the next chapter.)

· Groups like the Rainforest Action Network bring the spotlight to abuses in the forest products and oil industries and try to bring public pressure to bear to reduce and eliminate these abuses. Another highly effective organization is the Rainforest Alliance, which works with businesses to promote sustainable agriculture, forestry, and tourism. Like the Rainforest Alliance, the Forest Stewardship Council certifies that forest products have been harvested sustainably. Both the Rainforest Alliance and the Forest Stewardship Council have extensive relationships with forest product businesses, as well as with local governments and indigenous peoples.

· The World Resources Institute has programs like its Corporate Ecosystem Services Review and "Development through Enterprise," which seek to identify market opportunities and good business models for emerging economies, and ENVEST, which looks at the financial implications of risks and opportunities relative to environmental issues.

· Government agencies all over the world, like the EPA and Department of Energy in the United States, have extensive relationships with businesses, helping to promote research and development, working to institute best practices, and facilitating government loans, loan guarantees, and grants.

· Nonprofit consultants are involved as well. One prominent foundation, the Rocky Mountain Institute, headed by sustainability visionary Amory Lovins, has been working for years with corporate clients and partners as diverse as Boeing, Walmart, and the owners of the Empire State Building to streamline operations and maximize efficiency.

Ceres, among its many activities, also works with corporate boards to inform and improve their views on sustainability. The Sustainable Governance Forum on Climate Risk brings corporate leaders together with high-level academics (Yale), experienced risk managers (Marsh), and expert activists (Ceres) in small groups to look at the pitfalls in climate and pollution and the many advantages for business in a green approach. Mindy Lubber, the president of Ceres, refers to this as their "retail" work.

"Looking at climate and sustainability has to be integrated into the enterprise. In working with companies, we look at how to change governance. There needs to be a board committee looking at climate and sustainability, setting goals and analyzing those things. Why? Because the first duty of a fiduciary is to analyze risk — and climate is a risk."[79]

Beyond that? Ceres works with companies to integrate goal setting at the governance level. With many companies, Ceres works to make sure that hitting the goals is a determinant of executive bonuses. "Compensation is a very big piece of what we think a company interested in sustainability needs to be doing."[80]

Creating Opportunities

We have looked at climate risk and how companies, NGOs, and others are working to minimize them. Climate change does not, however, present only risks. The push to transition to clean tech presents considerable opportunity, not only to reshape how we do business and how we interact with our natural environment and use resources, but also to create economic growth and jobs.

The IPCC, and others, have a term for this: co-benefits. These are defined as "the benefits of policies implemented for various reasons at the same time, acknowledging that most policies designed to address greenhouse gas mitigation have other, often at least equally important, rationales (e.g., related to objectives of development, sustainability, and equity)."[81]

For instance, an analysis by the BlueGreen Alliance, a consortium of labor unions and environmental organizations, concluded that federal renewable energy policies, including a Renewable Electricity Standard (also known as a Renewable Portfolio Standard), could generate 850,000 jobs in manufacturing in the United States.[82] Among the alliance's members are the United Steelworkers, the Sierra Club, the Natural Resources Defense Council, the National Wildlife Federation, the Communications Workers of America, the Service Employees International Union, and the United Auto Workers.

Nicholas Stern wrote, "We have already embarked on what will be the most dynamic and creative energy and industrial revolution in our economic history: the transition to low-carbon growth. And this growth will be more energy-secure, safer, quieter, cleaner, and more biologically diverse."[83]

Every indication, as we have seen, is that we are well along in this transition.

Another economist wrote about big industrial concerns in Europe, "The large incumbents in Europe, which might have been considered technological laggards, have used green technology and sustainability as a core new element of growth."[84]

"Green products are the lead technology of the 21st century. This is the third wave of industrialization," said Peter Löscher, head of Siemens, a Fortune 500 company.[85]

President Obama flagged the job implications for American workers from new clean-tech deals at a speech in early 2011 at General Electric, another global industrial giant like Siemens with a steadily increasing emphasis on green manufacturing. Obama trumpeted the role of clean-tech manufacturing, for export and for domestic consumption, in producing thousands of jobs, and he also announced the creation of a new White House Council on Jobs and Competitiveness, chaired by GE's head, Jeff Immelt.[86]

Investments in clean energy plateaued somewhat in 2008 and 2009, owing to the global economic downturn — except in Asia (led by China and South Korea) — but 2010 gave every evidence of a return to the steep upward slope that had been very much the trend earlier in the decade. The figures for new money going into clean energy for 2010, including venture capital and corporate research and development, approached $250 billion worldwide, which is twice the figure recorded in 2006 and five times that from 2004. What were the principal drivers of this pop? China, European offshore wind projects and rooftop solar deployment, and R&D.[87]

With the world coming out of the recession and with growing confidence in the trends toward clean tech as a result of further progress on international action coming out of Cancún, as well as bilateral frameworks enabling deals, such as those that President Obama referenced in his speech at GE, things were looking up. One banker noted, "After the multiple headwinds that impacted climate investing in 2010, 2011 starts on a more optimistic note."[88]

Green Growth

As we have just noted, political leadership in the United States understands and supports the transition to clean tech because of its job-producing, economy-growing potential. So do other national and subnational leaders in the world. One of the most prosperous nations in the world, Abu Dhabi, rich from its oil, is putting a lot of its chips on renewable energy. In fact, Abu Dhabi is home to the Masdar initiative, one of the most ambitious clean-tech projects in the world. Masdar brings several elements under one umbrella, or in the case of the Arabian Desert, one parasol. When it is fully rolled out, Masdar will have a high-tech graduate research university, the Masdar Institute, being developed with MIT, one of the most prestigious engineering and science centers in the world. Masdar Power is developing major renewable-energy projects, and Masdar Carbon is seeking to reduce energy use and emissions through the commercialization of energy efficiency, waste heat recovery, and carbon capture and storage (CCS) projects. Masdar Capital is committed to making significant investments in clean tech and building effective, sustainable businesses. Perhaps most exciting of all the Masdar initiatives is Masdar City. It aims to run wholly on renewable energy, be highly energy efficient, with low waste and water use, and be a place where it will be a pleasure to live, work, and learn. It will house forty thousand residents and hundreds of businesses associated with the initiative. The Masdar Institute is already up and running in the first phase of the city.[89]

Joseph Stiglitz, a Nobel laureate in economics, and Nicholas Stern, in the teeth of the global recession of 2008 and 2009, encouraged countries to use "green stimulus" as a way to get their economies moving again. They reckoned that a robust insulation program in the United States "could create and sustain up to 100,000 jobs between 2009 and 2011, while saving the economy from \$1.4bn to \$3.1bn a year between 2012 and 2020." Further, they wrote: "This type of investment and those in green technology and infrastructure would not only provide a short-term stimulus but also improve the U.S. competitive position. As the world moves to a low-carbon economy, there will be a competitive advantage for those who embrace these technologies."[90] As we saw in Chapter 4, the Obama administration embraced this philosophy and devoted \$112 out of the \$787 billion total of the economic stimulus package to green purposes, about 14 percent.

The South Koreans, on the other hand, devoted the preponderance of their stimulus to clean-tech jobs and infrastructure. They want to create a million green jobs, switch all their lighting to highly efficient light-emitting diodes (LEDs), build nearly three thousand miles of bike expressways, increase mass transportation, and bring EVs onto the roads, among other projects. The president, Lee Myung-bak, a former construction industry top executive, embraces the same message that Stiglitz and Stern send: clean-tech advances will help promote competitiveness.[91]

It's the same message that President Obama touted in his speech on jobs and the economy at GE and numerous times before and since. It's the same message that the OECD has been signaling with its Green Growth Strategy. They analyze economic and environmental policies to find ways to spur eco-friendly growth. The OECD looks at jobs, taxes, trade, and investment in advising governments how to rev their economic engines.[92] The United Nations Environment Program has a similar project under way, its Green Economy Initiative. UNEP estimates that 15 percent of the \$3.1 trillion worth of stimulus worldwide was "green in nature."[93]

Reading the Tea Leaves

As we saw in Chapter 1, there has certainly been an explosion of coverage in the media on climate change and sustainability over the past couple of decades, and particularly in the past several years. Not the least of this coverage takes place on the business pages of newspapers and at websites and periodicals devoted to "green business."

The GreenBiz Group, for instance, operates four websites: ClimateBiz .com, GreenerBuildings.com, GreenerComputing.com, and GreenerDesign .com; holds conferences and "webinars" working with various partners; issues annual reports on the "State of Green Business" and also on the impact of green building; and runs a network of industry professionals.

Its annual "State of Green Business" report for 2011 is gung ho on the future of the marriage of sustainability and profitability. The report identifies a "steady march of progress" toward sustainability in businesses in 2010. It says that "a dramatic shift is occurring in business: Companies are thinking bigger and longer term about sustainability — a sea change from their otherwise notoriously incremental, short-term mindset."[94]

This contrasts, interestingly, with "the open hostility with which environmental protection is viewed by a swath of the political spectrum, at least in the U.S."[95] That "swath" of the spectrum in the United States is embodied in the modern Republican Party. Even as more and more Fortune 500 companies realize and act on the necessity of minimizing climate risk to their companies and embrace the opportunities in clean tech, the traditional party of business in America is fighting a vicious rear-guard action. The irony is thunderous.

Notwithstanding the unshakable convictions of the Grand Old Party, exacerbated and amplified by the recalcitrance of the Tea Party on climate and energy (which harks back to their ideological forebears in the Know Nothing Movement of the mid-1800s), the folks at the GreenBiz Group, among others, nevertheless see any number of important trends, including

- as consumers clamor for more sustainable products, manufacturers are increasingly answering the call;
- more and more companies are aiming to reduce their energy, water, and carbon footprints, as well as their toxic waste output;
- supply chains are being optimized to maximize efficiency and minimize greenhouse gases;
- with the advent of electric vehicles on the commercial scene and the continuing growth of hybrids, transportation is a greater priority, especially for companies with fleets;
- manufacturers and retailers are increasingly focused on managing

their processes and operations so as to aim for the cradle-to-cradle paradigm[96] of essentially zero waste.

Pushing the Edge of the Envelope

Procter and Gamble is the largest consumer products company in the world. P&G had $13.4 billion in profit on nearly $80 billion in revenue in 2010. Company leaders say publicly that they are committed to 100 percent renewable energy for their manufacturing, 100 percent recyclable materials for their packaging, and zero consumer and manufacturing waste.[97] If a company with that kind of juice has those kinds of commitments, then it is going to inevitably be giving a palpable push toward sustainability in the business world. Another consumer products giant, Unilever, had 2010 profits of $8.6 billion. It also has lofty goals to achieve by 2020: to cut its environmental footprint in half, to work with a billion people worldwide to improve their health and well-being, and to source 100 percent of the company's agricultural raw materials sustainably.[98] PepsiCo, with almost $6 billion in profits in 2010, is another company pushing hard to be a good actor on sustainability. Among its targets are to radically improve its water-use efficiency in production and to help provide access to clean water for billions, maximize recycling and reduce packaging weight, improve energy efficiency and apply proven sustainable practices in agriculture.[99]

These three companies, along with over sixty other consumer products giants and retailers, are members of the Sustainability Consortium. The group's goal is "to improve consumer product sustainability through all stages of a product's life cycle." They do this through establishing standards and tools for measuring product sustainability.[100]

When a corporation tries to paint its operations as clean, green, and sustainable without backing that up with real actions on the ground, we call that "greenwashing." Is what these corporations are doing greenwashing? To say that they are 100 percent sincere and devoted to the cause of sustainability would be naïve. Of course their shift in emphasis is, to a certain extent, motivated by improving their image, their "brand," as we say.

But to think that the transition in corporate values reflecting a much greater reliance on sustainability is wholly fictitious would be cynical and vastly underestimates the value of the changes that have been made. The influence of the corporate social responsibility movement during recent years has had an enormous influence on how these companies operate. Organizations like Ceres, the Rainforest Alliance, the Carbon Disclosure Project, and many others have effected a quantum shift in corporate behavior. These groups are watching and are intimately involved in many cases with the efforts of companies to produce positive environmental outcomes. The Sustainability Consortium, for example, has a membership that includes leading international environmental organizations like the Natural Resources Defense Council, the Environmental Defense Fund, and the World Wildlife Fund, not to mention key government bodies like the U.S. Environmental Protection Agency and the UK's Department for Environment, Food and Rural Affairs (Defra).

Returning to some salient cases of how corporations are making substantive changes and a very positive impact, Tesco, the fourth-largest retailer in the world, has, for its part, been on the bandwagon for a few years. Tesco, based in the United Kingdom, has 492,714 employees worldwide working in 5,380 stores. They have been reducing emissions and waste and opened the world's first zero-carbon supermarket, in the UK, in 2009.[101]

But the biggest kid on the block — and in fact the largest corporation in the world — is Walmart. Walmart had $14.3 billion in profits on $408 billion in revenues in 2009. That was about $120 billion more than number two, Royal Dutch Shell. Walmart's efforts on sustainability are particularly important for the simple reason that the company's vast size amplifies the impact of its programs. Walmart's stated aims are to be powered by 100 percent renewable energy, to be a zero-waste operation, and to reduce their climate impact all the way down their supply chain. They have, to these ends, been purchasing wind power for and installing solar photovoltaic arrays on their stores. They have been upgrading the energy efficiency of their existing stores and building new facilities with significantly higher efficiencies. They have been maximizing the average fuel efficiency in their fleets and been using more alternative fuels such as biodiesel, with an eye toward expanding their uptake of more-advanced fuels such as hydrogen, biodiesel from algae, and ethanol from cellulosic matter.[102]

One of their principal partners in this, the venerable and highly effective

Environmental Defense Fund, fully appreciates the value of having a giant like Walmart on board. EDF sees how strengthening the sustainability of the supply chain has "positive ripple effects around the globe," reaching tens of thousands of companies that might not otherwise be required by their local or national governments to reduce emissions and increase efficiencies.[103]

The unfortunate fact of rampant consumerism remains. How much and what we consume is a critically important factor in how we succeed or fail in confronting the climate crisis, and we will look more closely at this in Chapter 7, particularly in the context of our food. Walmart is, to be sure, one of the principal apostles of consumer culture. Without consumerism, Walmart would not be the giant corporation it is. In the meantime, though, Walmart's effort to reduce its greenhouse gas footprint is a crucially important, tangible development, as is the work of the other companies mentioned here.

What's in It for Us?

Yvo De Boer was the executive secretary of the United Nations Framework Convention on Climate Change from 2006 through 2010. In that capacity, he not only became intimately familiar with the ongoing international negotiations that had such dramatic peaks in Bali in 2007 and Copenhagen in 2009, but he also worked closely with businesses and NGOs to help effect positive change. He is now in the position of global adviser on climate change and sustainability for KPMG, business consultants with operations in 146 countries.

When it was announced that he was joining KPMG, he said: "Sustainability is high on the agenda of investors, companies and governments alike." Further, he noted "Although it is the role of governments to provide the necessary policy frameworks, I have always maintained that business will deliver the necessary innovation and solutions, providing the right conditions are created."[104]

De Boer underscored this perspective by saying "Companies are picking up on this and are not waiting for governments to act.... This is being driven by a broad agenda that relates to climate as a concern, but also concerns over energy prices, energy security, materials scarcity, and changing consumer behavior. That means that companies are looking at climate and sustainability through a new 'risk and opportunity lens,' and they're beginning to look at their operations and supply chains in a whole new way, and actively adjusting their operations."[105]

A recent KPMG report on corporate sustainability, done in cooperation with the Economist Intelligence Unit, surveyed nearly four hundred senior executives and found, among other things, that the main drivers are cost reduction, regulatory requirements, brand enhancement, and risk management. Some 72 percent of the executives from big companies (with revenues above $5 billion), think "that the benefits of investing in sustainability outweigh the costs." Perhaps counterintuitively—certainly if you believe what conservative politicians say, in the United States and elsewhere—the study found that "corporate lobbying activity is weighted towards tighter national and international rules, despite the recognition of a greater regulatory burden and increased operating costs." The report also gives a "snapshot of benefits" from sustainability as identified by the executives in the survey. These include risk mitigation, access to new markets, cost reduction, new products and services, better relationships with suppliers and clients, more ethical businesses, improved investor awareness, resource efficiency, and happier employees.[106] (Remember what architect Bob Fox said about his Bank of America Tower? BofA will realize $10 million annually in increased productivity because of the radically enhanced employee satisfaction from just working in such a wholesome environment.)

McKinsey & Company is another global business consultancy with a sharp focus on sustainability. It has nearly one hundred offices in over fifty countries. Its report on the opportunities for increased efficiency in nontransportation sectors of the American economy contained some arresting numbers: the potential to realize 23 percent savings in energy consumption by 2020, eliminating more than $1.2 trillion in waste, and thus abating 1.1 billion tons of greenhouse gas emissions annually.[107] Thankfully, the U.S. Department of Energy, states and cities, NGOs and the corporate community are becoming increasingly focused on realizing these gains.

Energy efficiency has certainly been on the agenda for a number of years in the United States. Efficiency gains have been responsible for a 65 percent to 75 percent increase in energy productivity since 1970.[108] A study from the Rocky Mountain Institute, the nonprofit consultancy founded and run by Amory Lovins, supports McKinsey's numbers on how much could be gained further: if the whole country could realize the efficiencies in the best-performing states, New York and California among them, the country could save 1.2 million gigawatt hours of electricity each year. That's 30 percent of Americans' electricity use.[109]

Putting it microcosmically, Dan Reicher, executive director of Stanford University's Steyer-Taylor Center for Energy Policy and Finance, says that by installing a new high-end, highly efficient refrigerator, his family is saving the equivalent cost of putting in a solar PV array. "When we do take the next step and put photovoltaics on the roof, it can be a much more downsized system than it otherwise would have been."[110] Amory Lovins would call the energy saved "negawatts" — energy that does not have to be generated because the need for it has been obviated by means of efficiency or conservation.

For the consumer, the perception that the upfront costs for more-efficient appliances such as refrigerators and air conditioners may be expensive could be a disincentive; but in reality, prices for major appliances have been getting lower, while efficiencies have been increasing. We are getting a lot more bang for the buck from our appliances than we used to do. A study looking at Japan, Australia, the United States, and Europe found "not only is the average energy consumption of appliances falling, but that they have also become cheaper." In the United States, for instance, the energy consumption by refrigerators and freezers fell by 60 percent from 1980 to 2001, while prices also decreased — by 40 percent.[111] Add to this the payback from lower energy costs, and the fact that tax incentives, rebates, and other supports are available to many consumers, and the economics soon become irresistible.

"Pollution Prevention Pays"

One of the pioneers in the movement among industrial concerns to capture the savings in reduced waste and energy was 3M — Minnesota Mining and Manufacturing — founded in 1902 and number 370 on the Fortune Global 500 list for 2010. In 1975, 3M introduced its 3P program, Pollution Prevention Pays, the idea being simple: using fewer expensive chemical resources, less energy, and generating fewer toxic wastes — which are extremely costly to manage properly — would net the company savings. In order to maximize the participation of its workforce in this effort, 3M gave cash awards to those of its engineers, scientists, and others who came up with new, less wasteful processes. In the thirty-five years of the program, over eight thousand projects have generated $1.4 billion in savings.[112]

Back in 1992, in summarizing the important themes to emerge from the Earth Summit in Rio de Janeiro, William Reilly, the head of the U.S. delegation, wrote, "A new dynamic may be at work in U.S. companies: They're finding ways to reduce pollution by redesigning processes, improving efficiency, and cutting the costs of raw materials, disposal, and potential liability."[113]

Green Is the New Gold

Nearly twenty years later, he has seen a lot of progress. Reilly has been on the board of DuPont since 1993 and has chaired the Environmental Policy Committee of the board since that time. DuPont employs fifty-eight thousand people, with operations in about eighty countries, and makes all kinds of high-tech products for agriculture, transportation, communications, and electronics, among others sectors. Reilly has witnessed the enormous commitment that DuPont has made over time to clean up its inactive hazardous waste sites — under pressure, to be sure, from federal and state agencies. DuPont has also done a lot of work to reengineer its processes, as 3M and other chemical manufacturers have, to reduce the production of toxic byproducts and to save energy.

But a few years ago DuPont went to a new level. It began to make "market-facing commitments." In Reilly's words, "This is a real revolution. They took their portfolio of businesses apart and they said, 'Let's adjust this company to an expectation of a future of a greater population, of increasingly greater need for food and nutrition, for growing scarcity of commodities and stresses on the environment, for water scarcity associated with climate change (in many places), and adapt the business itself to those realities, but not in a defensive way, rather in a proactive way that will earn us money from the trends that we foresee.'"[114] DuPont is realizing this path in a host of ways, including developing new seeds that can thrive under different temperature conditions and with higher salinity in the soil, by making new fibers from cellulosic crops like switchgrass, by creating advanced biofuels, improving fuel cell technology, and by manufacturing building materials with a very low environmental footprint.

United Technologies (UTC), a company with $53 billion in annual revenues, makes big-ticket items: helicopters, aircraft engines, heating and air conditioning systems, fuel cells. Its Otis division has 2.3 million elevators and escalators in operation around the world. Like DuPont and the others, UTC is also committed not only to reducing waste and greenhouse gases, but also to building on the many clean-tech market opportunities available to its businesses. Its fuel cell, HVAC, and elevator divisions, for instance, are working to increase the ability of their products to enhance the efforts of architects and developers building sustainable buildings. This carries over to their work in the nonprofit sector as well. UTC co-chairs the World Business Council on Sustainable Development's task force on Energy Efficiency in Buildings (EEB). The "manifesto" for the task force's project calls for rigorous reporting on energy use in buildings and a number of programs to reduce energy use.

For Ford Motor Company, there is the obvious incentive to design, manufacture, and sell the full range of hybrid electric, plug-in hybrid electric, and all-electric vehicles for which demand is growing quickly. Beyond this, though, Ford has taken a big step in moving toward a highly innovative approach to clean tech in the manufacturing process itself. In 1999, Bill Ford Jr. commissioned visionary green architect William McDonough to rebuild one of his company's landmark plants. The Rouge River facility had once employed over a hundred thousand workers, but it had fallen into disrepair and was mired in many acres of highly toxic contamination from decades of indifference to the environment and the health and safety of its workers. McDonough devised a $2 billion plan to not only clean up the wastes using an effective but low-cost process of "phytoremediation" (the use of natural plants to rid the ground of contamination), but to build a million-square-foot truck manufacturing facility with a ten-acre green roof, advanced natural stormwater management systems, considerable planting to advance the "biophilic" character of the workplace, and a large reliance on renewable energy. Today, the Rouge River Center houses one of Ford's biggest truck factories and employs thousands of workers.

Siemens, another global industrial giant, has put a lot of emphasis on its energy and transportation divisions in recent years. The payoffs have reflected the company's confidence in these sectors. Chief executive Peter Löscher put it this way in his speech to a recent annual general meeting:

Our "Green Portfolio" is a triple win: for the environment, for society, and for you, our owners. "Green" is worth it. For example, last fiscal year the answers we have in our environmental portfolio helped our customers reduce their CO_2 emissions by about 270 million metric tons worldwide. That's equivalent to the annual emissions of Hong Kong, London, New York, Tokyo, Delhi and Singapore combined! Our environmental portfolio generated revenue of some 28 billion euros — meaning that today we have already outperformed our initial goal for fiscal 2011. We now plan to increase our "green" revenue to more than 40 billion euros by 2014.[115]

From high-speed intercity trains and subway systems, from massive gas turbines and both onshore and offshore wind power equipment, to sophisticated smart grid technology and high-tech and highly efficient transmission cables, to various building technologies including advanced lighting solutions, and with many other applications building its business, Siemens has fully embraced the new industrial revolution.

"I Have Seen the Future and It Works"

That's what Lincoln Steffens said, famously, in 1921 about Soviet communism. After ten years' time, however, he realized it didn't work after all. In any event, it's a great line, and it is highly applicable today to what we are experiencing.

The great renewable-energy seer and activist Herman Scheer described our postindustrial economies as being in transition from being based on fuels to being "technology-based." If we are able to successfully effect this change, and rapidly emerging economies are able to "leapfrog" to this paradigm, then the pressures on the environment, public health, and the climate system will be radically diminished. That major corporations from sectors as diverse as manufacturing, insurance, retail, and finance, and even some of the biggest energy and fuel companies, fully recognize and embrace the opportunities in this transformation — and the risks in not making it — then we have considerable hope for the future.

Speaking of these innovations and advances in manufacturing and elsewhere in the world of global business, William Reilly noted: "It's always fascinating to me that this can be true among some of our most sophisticated companies and yet our political leadership seems to have not caught on."[116] While that may be true for some top politicians in the United States, even at this late date, it is not at all the norm for most of the world's top policy makers and for those making change at the national, regional, and local levels, as we will see, abundantly, in the next chapter.

planet green

THE POLICY, POLITICS, AND PRACTICE REVOLUTION

Fossil Fuels:
Will the Rapidly Developing Economies Swamp the Ship?

Global emissions of carbon dioxide in 2009, not counting land-use changes, were 30.303 billion tons (see figure 6.1). Of this, China was responsible for 25 percent, and the United States, 18 percent. The good news is that the United States has been reducing its output. The bad news is that China's emissions have been increasing. Other good news is that the United States pledged at the UN conference in Copenhagen in 2009 to reduce its overall greenhouse gas emissions 17 percent from its 2005 levels by 2020. China, as noted in Chapter 4, made this commitment:

> China will endeavor to lower its carbon dioxide emissions per unit of GDP by 40–45 percent by 2020 compared to the 2005 level, increase the share of non-fossil fuels in primary energy consumption to around 15 percent by 2020 and increase forest coverage by 40 million hectares and forest stock volume by 1.3 billion cubic meters by 2020 from the 2005 levels.[1]

When he was in Copenhagen, President Obama trumpeted the importance of China's, and India's, acceptance, for the first time, of "their responsibility to take action to confront the threat of climate change."[2]

Yet although the United States, the European Union, and some others are working very hard on some fronts to reduce their emissions, fossil fuel use is still growing worldwide, coal being the biggest contributor to the pool. If some developed economies are working hard to get their footprints down, will their reductions, in the face of rapidly growing emisisons from the developing economies, be enough to adequately mitigate climate change?

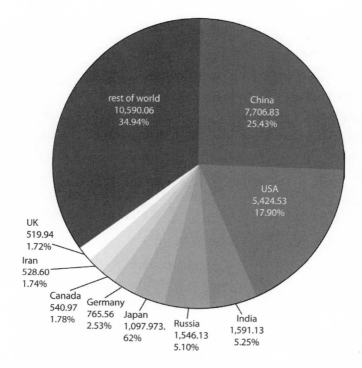

rest of world
10,590.06
34.94%

China
7,706.83
25.43%

USA
5,424.53
17.90%

UK
519.94
1.72%

Iran
528.60
1.74%

Canada
540.97
1.78%

Germany
765.56
2.53%

Japan
1,097.973.
62%

Russia
1,546.13
5.10%

India
1,591.13
5.25%

6.1 Emissions of CO_2 in millions of tons, and percentages of total. Graphic by the author and Lorraine Simonello, based on data from the Energy Information Administration of the U.S. Department of Energy

Some figures for the first decade of the twenty-first century:

· While the United States decreased its CO_2 emissions from coal from 2.155 billion tons in 2000 to 1.882 billion in 2009 (-12.67 percent),
· Japanese emissions increased slightly, from .360 billion tons to .389 billion tons (+8 percent),
· in China emissions went from 2.157 billion tons of CO_2 in 2000 — nearly the same as the United States at that time — to 6.477 billion tons in 2009 (+200 percent),
· in India, emissions increased from .673 billion tons to 1.09 billion tons (+62 percent).

The world's total carbon dioxide emissions from coal increased 54.75 percent in those ten years — from 8.655 billion tons to 13.394 billion tons.

Of that increase of 4.739 billion tons, China was responsible for 4.32 billion tons—fully 91 percent. India was responsible for much of the rest.

Similarly:

- U.S. emissions of carbon dioxide from oil consumption went from 2.461 billion tons in 2000 to 2.319 billion in 2009 (-5.77 percent),
- Japan's emissions from oil decreased from .678 billion tons to .511 billion (-24.63 percent),
- while in China, CO_2 emissions from oil increased from .643 billion tons to 1.06 billion tons (+65 percent),
- and in India, the figure went from .282 billion tons to .393 billion (+39.43 percent)[3]

These last two rapidly developing economies are, not incidentally, the two most populous countries in the world by far. Of the nearly seven billion people in the world in 2011, China had about 1.337 billion and India about 1.189. Together their 2.5 billion people comprise 36 percent of the world's total population.[4]

The carbon dioxide from coal and oil combustion from these two countries clearly is a major problem for the world in addressing the climate crisis—probably the single-largest problem of all. Unfortunately, there are other climate and pollution implications associated with China and India.

In India and China, the burning of biomass, as well as coal, for cooking and heat, by hundreds of millions of villagers, and the combustion of coal for electricity and other industrial purposes generate tremendous amounts of soot, or black carbon, as well as many other pollutants, including carbon monoxide. The devastating health impacts from burning biomass (wood, straw, and dung) in open hearths or coal in primitive furnaces and ovens in small, badly ventilated homes has been well known and documented for some time. In India and China this practice leads to hundreds of thousands of deaths annually—pneumonia being the leading cause among children, and bronchitis, emphysema, and lung cancer for adults. In China, around 80 percent of the population relies on solid fuel; in India, it's 74 percent.[5]

But black carbon also has a pernicious effect on the climate through its influence on snow and ice. Black carbon induces warming by darkening bright reflective surfaces that normally bounce solar radiation back to space and thus increasing those surfaces' subsequent absorption of incoming sunlight. In the

Himalayas, where the glaciers are diminishing at astonishing rates, there has been a measurable deposition of black carbon since at least the early 1950s, with strong increases since 1990. One leading researcher, Veerabhadran Ramanathan, and his colleagues have identified a very strong correlation between the melting of the glaciers and black carbon deposition. Half the melt is a consequence of the soot and half from temperature increases due to warming from the buildup of other greenhouse gases.[6] One expert explains it this way: "Black carbon from India is more immediately deposited on Himalayan glaciers via warm, moisture-laden, southerly monsoon winds that sweep it up onto the Tibetan Plateau." As black carbon builds up, it turns the glaciers into "giant collectors of solar heat."[7] A similar level of impact on the Arctic is occurring as a result of black carbon.

Conventional air pollutants from power plants, industrial facilities, and the rising tide of motor vehicles in India and China are another enormous burden on these populations, and beyond. Acid rain, for instance, costs China $4.5 billion annually in crop losses and a billion more in damage to buildings. China is the largest source of sulfur dioxide (SO_2) emissions in the world. SO_2 is the principal precursor of acid rain.[8]

Sulfur dioxide is also an air pollutant that causes considerable health damage, as does particulate matter from the incomplete combustion of fossil fuels. Ground-level ozone, a potent greenhouse gas (and increasingly being recognized for its impact in this regard), causes considerable problems for people, especially for children and seniors with sensitive lung tissue. Carbon monoxide, largely emitted from surface transportation, is another potent air pollutant. All these air pollutants are produced in one degree or another as a consequence of burning fossil fuels. Chinese cities have some of the worst air pollution in the world, far exceeding, on average, the worst of American cities (San Diego, Los Angeles, and Houston) for particulates, a very dangerous substance. In South Asia, cities with highly elevated levels of particulates included Kathmandu, Dhaka, New Delhi, Kolkata (Calcutta), Mumbai (Bombay), and Colombo.[9] In 2003, cities accounting for 58 percent of China's population reported particulates at more than twice the levels of the U.S. average.[10]

The government of the People's Republic of China has traditionally been reticent to acknowledge sweeping ills like massive pollution and its health impacts. That appears to be changing. The environment minister noted the dire situation early in 2011: "In China's thousands of years of civilization, the conflict between humankind and nature has never been as serious as it is today. The depletion, deterioration and exhaustion of resources and the worsening ecological environment have become bottlenecks and grave impediments to the nation's economic and social development."[11]

Smoother Sailing on the Horizon?

It is abundantly clear that we are not going to confront the realities of climate change successfully if China, India, and the other rapidly developing economies, among them Indonesia, Malaysia, Brazil, South Africa, and Mexico, don't get a handle on their greenhouse gases. The implications for the well-being of billions of people in these countries are also plain: both the direct effects of fossil fuel combustion and industrial production manifested in water and air pollution, and the indirect effects from warming and other climate changes on health, food, and water supplies, as well as on economic progress, are being felt now and will accelerate and intensify if measures are not taken.

Thankfully, many governments, multilateral institutions, NGOs, and individuals have long since recognized these truths and have been acting to address them. Many more are getting on board to help. This is true not only for the increasingly energy-hungry countries we have looked at here, but for the well-developed economies as well. Even in the absence of an overarching global regime in which countries agree to substantially reduce their greenhouse gas output, we are seeing progress.

The Greening of China

China, as noted, made a commitment in Copenhagen to reduce its "carbon intensity." Premier Wen Jiabao reiterated and refined this goal in early 2011, pledging to cut carbon intensity around 17 percent from 2011 levels by the end of 2015.[12] China measured 2.219 metric tons of CO_2 per thousand U.S. dollars of economic output for the year 2009, up from 2.011 metric tons in 2000. By contrast, the number for the United States in 2009 was .421 metric tons, down from .522 metric tons in 2000, while the overall global number has increased from .598 metric tons in 2000 to .615 metric tons in 2009.[13]

But reducing carbon intensity does not necessarily mean cutting gross emissions, because the overall output of the economy may — and in China's case, certainly will — continue to increase. The question is how much and how fast can China reduce its emissions growth and hopefully, eventually, lower its greenhouse gas output.

An expert group at one of the U.S. Department of Energy's top labs devotes itself solely to looking at China and energy. A comprehensive study that the group mounted looks at projections for energy use and carbon emissions through 2050. The general assumptions for China have been largely along the lines that energy use will continue to grow by leaps and bounds through at least the end of the century and that emissions will grow at roughly the same rate. The DOE's experts, however, have modeled several scenarios and find it is distinctly possible "that this will not necessarily be the case." Why? For two principal reasons: (a) by 2030, China will have reached a saturation point with respect to energy use by the economy in general and consumers in particular (aided by slowing population growth), and (b) energy efficiency initiatives and a trend toward a "decarbonized power supply" owing to the deployment of renewables may well significantly advance.[14]

It is critical for the world's climate system that China achieve progress on energy use and the decarbonization of its energy economy. It is instructive therefore to look at some of the extraordinary initiatives on clean tech that China has been launching.

In 2009, China was the world's biggest producer of hydroelectricity, with 197 GW of installed capacity. (Total installed capacity for all electric power was nearly 900 GW.) The government is seeking to nearly double hydroelectric capacity by 2020.[15] Hydroelectric dams are controversial, and the Three Gorges project, perhaps the most controversial in the world, displaced over a million people and will be the biggest facility in the world as of 2011, generating 22.5 GW. One environmental concern regarding Three Gorges is that the massive area that it inundated was home to hundreds of mines, waste dumps, and industrial plants that have made the water highly polluted. Another concern is that erosion downstream from the dam is causing landslides.[16] And although dams generate carbon-free electricity, they also generate methane, a greenhouse gas over twenty times more powerful than carbon dioxide. Methane, as well as carbon dioxide and nitrous oxide, come from decaying vegetation and soils and may be responsible for as much as 5 percent of the world's greenhouse gas emissions annually, according to one major study published in 2007.[17] Thankfully, a high percentage of these greenhouse gas emissions can be mitigated by capturing them at the turbines, an approach that the authors of the study recommend.

In any event, powering China from hydroelectricity more and coal less is a consummation devoutly to be wished.

Wind power has seen stunning growth in China in the past decade and will continue to accelerate. By the end of 2010, installed capacity was a little less than 42.3 GW, nearly 22 percent of the world total of 194.4 gigawatts.[18] The aim for China, though, is to hit 200 GW by 2020, displacing 440 million tons of carbon dioxide from coal combustion and substantially reducing air pollution. Of this, 32.8 GW is expected to be developed offshore.[19]

Concern regarding air pollution is one driver of this explosion of wind power development, as is a recognition that climate change is a reality and that greenhouse gases must be curtailed. China is facilitating this transition by giving wind power priority for grid connections, requiring grid operators to take a certain set amount of renewable energy, and by offering premiums to help offset the higher initial costs of wind installations and also by providing a feed-in tariff over the long term.[20] The renewable energy feed-in tariff, as we saw in Chapter 2, has been one of the principal supports for the expansion of the German and the Spanish renewable energy industries.

For solar power, China's PV capacity remains relatively small, with 310 MW in 2009, about 2 percent of the global installed capacity. By contrast, China had 135 million square meters (about fifty-two square miles) of solar water heaters by 2009, accounting for 70 to 80 percent of the global market.[21] Noting as we have the terrible health implications of power and heat from coal and biomass, the continuing move to solar is another giant leap for the Chinese people.

One more avenue for the tapping of the sun's limitless power is the deployment of concentrated solar power (CSP). Unlike solar photovoltaic, which captures the sun's energy and turns it directly into electricity, CSP uses the sun to create a superheated medium, like water or molten salt, that will create steam to drive a turbine, exactly as a fossil-fuel or nuclear power source does. The turbine in turn drives an electrical generator. An American company, eSolar, is in partnership with a Chinese company, Penglai Electric, to develop 2 GW of electricity over the next ten years.[22]

One very prominent analyst and activist, Joe Romm, calls CSP "the technology that will save humanity."[23] As noted in Chapter 2, CSP is one of the two core technologies for Desertec, along with wind power. Also previously noted is that at least one study says if we are hard-charging on CSP, we "could meet up to 7% of the world's power needs by 2030 and fully one quarter by 2050."[24]

There is yet another avenue for China to pursue in revving up its renewable energy portfolio: geothermal. This is still an underused resource all over the world, but the Chinese are partnering with the Icelanders to develop it.[25]

With China being the manufacturing powerhouse that it is, you would not be surprised to learn that it has been seizing momentum in building and exporting key renewable energy components. In 2009, China had the largest market share in the world in PV, with 40 percent, most of that going to Europe. For solar water heaters, the Chinese have an astounding 90 percent of market share.[26] In 2010, four of the world's top ten wind turbine manufacturers were Chinese.[27]

The venerable Worldwatch Institute says that "China's rapid rise to global leadership in clean energy is rooted in an unusual level of cooperation between government and industry, with the government providing a broad range of incentives that have led to the creation of renewable energy industrial bases nationwide."[28] This has led, perhaps inevitably, to some international trade tensions. The United Steelworkers petitioned the Obama administration to bring a case at the World Trade Organization because China was unfairly subsidizing its wind power equipment exports. On December 22, 2010, the United States did precisely that. U.S. trade representative Ron Kirk said at the time: "Import substitution subsidies are particularly harmful and inherently trade distorting, which is why they are expressly prohibited under WTO rules."[29] A coalition of American solar manufacturers has also filed complaints with the U.S. government against the Chinese for "dumping" of equipment on the American market, supported by the Chinese government's "arsenal of land grants, contract awards, trade barriers, financing breaks and supply-chain subsidies to advance its pricing and export aggression."[30]

Looked at another way, though, the exponential growth of China's renewable export industries, in the words of the Worldwatch Institute, "[has] implications that go well beyond China. As the country's skills in efficient, low-cost manufacturing are brought to clean energy industries, this could widen the energy options for the world as a whole."[31]

India's Renewable Energy Path

India generated a total of 529.12 billion kilowatt hours in 2000 and 835.27 by 2009, a 58 percent increase. (The United States, by contrast, generated 4,000 billion kWh in 2009.) For renewables, India went from 76.60 in 2000 to 121.78 in 2009, a 59 percent increase.[32]

Only two-thirds of the Indian people have access to electrical power, and a large majority of the population remains in the underserved, primarily rural areas. There is, then, a big gap in electricity supply yet to be made up. It is perceived that renewables can help make up the existing deficit and be the key component in ramping up Indian electrical supply to the 350 GW of demand projected for ten years hence, more than triple the existing supply.[33]

In 2003, the national government created some new carrots and sticks, including requiring grid operators to open their transmission lines and distribution systems to generators of renewable energy, requiring minimum obligations for the purchase of power from these developers, tax exemptions, and a premium price for their power. The government has also been extremely supportive in the critical area of project finance. As of the end of 2009, the Indian Renewable Energy Development Agency had financed nearly two thousand projects with loan commitments of over $2.4 billion. Other government agencies, plus multilateral development banks like the World Bank and the Asian Development Bank, have also helped underwrite renewable installations in India, as has the Clean Development Mechanism (CDM) operating under the auspices of the UN Framework Convention on Climate Change.[34]

One key study identifies all the obvious reasons that India wants to develop renewables:

- it has abundant solar irradiation, enormous potential for biomass production, wind resources both onshore and off, and thousands of miles of coastline and rivers for both conventional and advanced hydropower;
- renewables provide energy security;
- renewables protect against energy price volatility;
- off-grid, distributed generation can meet the needs of hundreds of millions of rural Indians;
- renewable power can serve India's poor;
- using renewables will help achieve India's climate change goals;
- India wants to become a leading player in clean tech, including renewable energy.[35]

One of the top renewable energy and energy efficiency analysts and advocates in the world, Daniel Kammen, was appointed the World Bank's chief technical specialist in these areas in the fall of 2010. Kammen notes that India has 150 GW of renewable energy generating potential. The challenge for India also lies in expanding and improving its grid, including bringing the one hundred thousand non-electrified villages on line and/or into a system of locally generated renewable energy.[36] To that end, the World Bank approved a $1 billion loan in 2009.

Kammen refers to another major study, from the World Bank, that identifies how the government can achieve its target of 55 GW from renewables by 2022: doubling wind power, quadrupling its existing small hydro output, optimizing cogeneration, increasing power from biomass by five or six times, and realizing the ambitious goals of the Jawaharlal Nehru National Solar Mission (JNNSM) to go from a mere 9 MW in 2010 to 20 GW by 2022.[37]

The first phase of the JNNSM is well under way, with $964 million approved to finance 1 GW of grid-connected large solar plants, 100 MW of rooftop and small solar plants, and 200 MW of off-grid thermal and PV.[38] Beyond this, the Desertec Foundation, concentrated primarily on laying the foundations for massive solar and wind in the Sahara and Middle East, is also looking at India. Desertec reckons that the Thar desert in Rajasthan has enough insolation — exposure to the sun — to power the whole of South Asia.[39]

What is the economic viability of a substantially renewable power base for India? According to the World Bank, the answer is: perfectly viable, owing to a number of factors. Among these are that the price of coal and oil have continued to trend upward, and the costs of capital equipment for renewables have been trending steadily downward.[40]

Unlike China, India is not a manufacturing powerhouse. Its industrial sector accounted for 28.6 percent of GDP in 2010, with 16.1 percent of GDP in agriculture and 55.3 percent in services. China's industry, by contrast, was 46.8 percent of its economy, with 9.6 percent in farming, and services at 43.6 percent. India's industrial production generated $348 billion in 2009, and China's, $2.3 trillion.[41]

However, India is serious about building a presence in clean energy manufacturing. By 2010, for instance, there were seventeen companies in India producing wind turbines capable of generating 7.5 GW. That production capacity may hit 17.5 GW by 2013. An Indian company, Suzlon, has a global presence and is one of the top manufacturers in the world. Other companies with a global presence producing wind power equipment in India include Gamesa, Siemens, and Vestas, three of the giants.[42] The Jawaharlal Nehru National Solar Mission has made a commitment to R&D and support for manufacturing of solar energy products as well.[43]

U.S. secretary of state Hillary Clinton encouraged India's venture into clean tech during her visit there in 2009. She talked about "leapfrogging."* Clinton said:

> And just as India went, from a few years ago, having very few telephones to now having more than 500 million mostly cell phones by leapfrogging over the infrastructure that we built for telephone service, we believe India is innovative and entrepreneurial enough to

---

* "'Leapfrogging' is the notion that areas which have poorly developed technology or economic bases can move themselves forward rapidly through the adoption of modern systems *without going through intermediary steps*" — from Worldchanging, "Leapfrog 101," http://www.world changing.com/archives/001743.html.

figure out how to deal with climate change while continuing to lift people out of poverty and develop at a rapid rate.

Obviously, these decisions are up to the people of India, but the private sector, based on our conversations, is looking for economic opportunities in clean energy and looking for ways to figure out how to move toward low-carbon energy production. So we're going to be engaged in these conversations.[44]

Backing up those words with action and money, when President Obama went to India in November of 2010, he and Prime Minister Manmohan Singh announced several high-profile initiatives, including a Joint Clean Energy Research and Development Center that will mobilize up to $100 million in public and private-sector funding over five years, as well as the Overseas Private Investment Corporation, providing $100 million in financing for the $300 million Global Environment Fund's South Asia Energy Fund.[45]

There is a lot of upside for India in renewable energy and energy efficiency — for manufacturers, investors, labor, and for the citizens of what will be the most-populous country in the world by 2030, and which is now the largest democracy by far.

Big Economies: Big Progress and Plans

India and China are obviously key to any hope we have of successfully slowing down and eventually reversing the course of greenhouse gas loading to the climate system. Other countries, though, have been rising to meet the challenge — and recognizing the many opportunities available in clean tech.

Green Tigers: East Asia

South Korea, for example, has put an extraordinary emphasis on the greening of its economy for some time now. In 2009, when economic stimulus packages were being put into place throughout the world to help avert a global depression in the wake of the financial crisis, South Korea devoted the greatest part of the pie, by far, to green initiatives. Nearly 80 percent of its stimulus package went to projects like those for increasing the use of hybrid vehicles, expanding forest coverage, and expanding renewable energy and energy efficiency.[46] The Koreans committed themselves to a "Green New Deal Job Creation Plan" to advance these projects, aiming to create nearly a million jobs in the process.[47] Among the concrete initiatives to which the Koreans are devoting money and political will is the smart grid, the innovative approach to electricity transmission and distribution that we examined in Chapter 3. It announced an initial expenditure of $7.18 billion early in 2011 with the aim of having the entire country on board by 2030. The smart grid, as we have seen, enormously enhances energy efficiency through demand-side management of power and radically reduces transmission and distribution losses, as well as greatly increasing the potential for the uptake of distributed and centrally generated renewable energy.[48]

The South Koreans have been so focused on this strategy of "green growth"—even having been credited with creating that term—that other countries and key multilateral agencies like the Organisation for Economic Co-operation and Development are emulating their initiative. Korea is the home of the Global Green Growth Institute (GGGI), founded by Nicholas Stern and Dr. Han Seung-soo, the prime minister of South Korea when the Green New Deal was launched. The GGGI aims to support primarily emerging economies in this effort.[49]

Taiwan, ranked nineteenth in the world in GDP for 2010, has taken the renewable energy bull by the horns too. Although Taiwan has not brought much solar power into play yet domestically, it, ironically, has been a world leader in the development and manufacturing of photovoltaic cells, with 3 GW of cell production in 2010, taking over the number-two spot from Japan. (China and Taiwan together accounted for nearly two-thirds of the global production of PV in 2010.) This is all going to change as Taiwan aims to boost its PV use from 19 MW in 2010 to 2 GW by 2025. In total, the country wants to double the share of renewables in its power from 8 percent—mostly hydropower—in 2010 to 16 percent by 2025. Taiwan is also making a push on offshore wind and ocean energy.[50]

Japan is the tenth-largest country in population, with the third-biggest economy. It also contributed over a billion tons a year of carbon dioxide from energy consumption in 2009, down 8.61 percent from 2000 but a formidable problem for the climate system nevertheless.[51] The speed and trajectory of Japan's path to a low-carbon future are, obviously, of paramount importance.

Japan, of course, was where the Kyoto Protocol was negotiated in 1997. It has consistently tried to advance the cause of climate change mitigation as an early signatory of the protocol, by being a leading developer of clean tech, very much including high-MPG hybrid cars, and by its extraordinary performance in financing mitigation and adaptation through various global climate funds. It has pledged over $16.5 billion as of mid-2011, far eclipsing the $6 billion the UK has pledged and $2.25 billion from the United States.[52]

As of mid-2010, Japan had 24 GW of renewable energy capacity, most of it from biomass. The Ministry of Economy, Trade, and Industry, a historically extremely powerful government agency, seeks to add 32 to 35 GW of renewable capacity in the next ten years.[53] Japan ranks in the middle among the thirty major countries ranked by Ernst & Young in their annual look at how attractive these countries are for renewable energy development and investment. One aspect of the rating is how much has been earmarked for the smart grid. In the case of Japan, that is $7.3 billion over the next ten years.[54]

Japan has led the world in the production of hybrid vehicles. The first and still the foremost of these, the Toyota Prius, surpassed two million in total sales worldwide in September 2010.[55] The Japanese auto industry is led by Toyota, but includes Honda, Nissan, and Suzuki all in the top ten manufacturers in 2009. Nissan, in particular, has made a very big bet on electric vehicles. The Renault-Nissan alliance chief executive, Carlos Ghosn, is predicting global sales of 500,000 to one million vehicles per year by mid-decade. Honda, also very active in selling hybrids and electric vehicles, is producing a fuel-cell car as well, the FCX Clarity.

The economies of Japan, South Korea, and Taiwan all have enormous influence. That they are devoting growing effort to advancing clean tech and fighting greenhouse gases is one more hopeful indicator.

Europe

We have seen how thoroughly Europe has embraced the promise of the push toward a clean-tech future, driven by the specter of climate change, the recognition of the value of energy security, and the abundance of economic development opportunities. The "20/20/20" program to radically reduce GHGs, increase the share of renewable energy, and improve energy efficiency is the law of the land in the twenty-seven states of the European Union. There is even a recommendation from the EU Commission to up the ante to a 25 percent emission cut by 2020.[56] We have also seen how Europe has embraced the program of environmental trading that was one of the "flexible mechanisms" in the Kyoto Protocol.

Europe's carbon dioxide emissions* have diminished modestly so far, from 4.546 billion tons in 1990 to 4.307 billion tons in 2009, a 5.26 percent decrease. CO_2 emissions, though, have decreased 8.69 percent from their peak of 4.717 billion tons in 2006. This does not include other greenhouse gases, of course, like methane and nitrous oxide, but it certainly indicates a positive trend overall.

Some states are doing better than others. The unified German state generated 928.95 million tons of carbon dioxide in 1991, but only 765.56 by 2009—a resounding 17.59 percent decrease. The UK went down from its peak of 609.98 in 1991 to 519.94 by 2009, a 14.76 percent decline. Some states, however, primarily because of rapid economic development, have increased their emissions. Spain, for instance, has gone from 237 million tons in 1991 to 329.86 in 2009, up 39.18 percent. Spain is, however, down almost 15 percent from its peak in 2007. Turkey, similarly, went from 137.29 million tons in 1991 to 253.06 in 2009, up 84.33 percent, but trending down from its peak of 280.19 million tons of CO_2 in 2007, a 9.68 percent improvement since that time.[57]

The EU states, though, are thinking big.

- The United Kingdom, for example, has leased offshore sites to wind developers who could well be providing 25 percent of that nation's electricity by 2020.
- Ten nations on or near the North Sea have entered into a compact to build a "supergrid" to maximize the renewable energy potential in the region, with companies like Parsons Brinckerhoff and Siemens involved in the process.
- The EU is mandating the movement toward zero-carbon buildings

---

* Europe for these purposes counts all the states of the EU, plus those that are not EU members (but not the former Soviet states).

and is providing the regulatory infrastructure and financial support to get it done. The goal is for new buildings and major renovations to be zero or close to zero energy by 2021. France is looking for all new construction to actually produce energy by 2020.[58]

· The smart grid is an integral part of the European vision for a low- to zero-carbon profile for society. The estimate for investment in the smart grid is $80 billion through 2020, and 240 million smart meters are the target for installation by then.[59]

· Distributed generation of energy is a concept and practice being increasingly deployed in Europe. Germany, for instance, installed 3 GW of photovoltaic power in 2009, 80 percent on rooftops. (That 3 GW for 2009 is three times the *total* installed capacity in the entire United States.) Of the facilities comprising Germany's 27 GW of wind power, 90 percent are 20 MW or smaller. Overall, Germany has a 43 gigawatt renewable energy market, with individuals accounting for 42 percent, farmers 9 percent, and utilities only 13 percent.[60]

The EU is aiming to reduce its 1990 output of greenhouse gas emissions by 80 percent by 2050. A lot of the action in reaching toward that goal is taking place at the subnational level, and we will return to that here.

Does the Bell Toll for Coal?

We have seen how in some of the critically important economies, from China and the rest of East Asia to India to Europe, change is in the air. Some of what we have examined at the national level indicates a seemingly inexorable trajectory toward a much healthier, greener planet. We have also had considerable insight into the international and national political breakthroughs, technological advances, and attention to clean tech by business and finance that have been moving us toward a newer world.

There are other developments indicating that, in a number of different ways and in many venues across the world, coal's role in energy and industrial production is being slowed and, in some cases, palpably reversed.

Texas, for instance, is the leading American state for installed wind power capacity, with over 10 gigawatts.[61] That is equal to the output of twenty

500 MW coal-fired power plants. That is a particularly interesting comparison, because Texas was the site in 2007 of a major breakthrough on turning back the tide of coal. TXU, a major regional utility, had proposed to build eleven new plants. A coalition of farmers, ranchers, environmentalists, and municipal leadership from some of Texas's biggest cities — Houston, Dallas, Fort Worth, Arlington, El Paso, and others — banded together to fight the plants. A tremendous groundswell of opposition built, and a dramatic climax ensued: TPG, a private investment firm, with William K. Reilly as its lead negotiator — the very same William Reilly who headed the American delegation to the Earth Summit — hammered out a deal with Fred Krupp, president of the Environmental Defense Fund, a leading opponent of the new coal plants, to eliminate eight of these and to allocate $400 million for energy efficiency. With the terms in place, TPG then bought TXU. Environmentalists characterized this historic deal as a "game changer."[62]

Later that same year, the secretary of the Kansas Department of Health and Environment, Roderick Bremby, blocked two 700 MW coal-fired power plants, saying "I believe it would be irresponsible to ignore emerging information about the contribution of carbon dioxide and other greenhouse gases to climate change and the potential harm to our environment and health if we do nothing."[63] He was backed up every step of the way by Governor Kathleen Sebelius, and a particularly nasty political and public relations campaign ensued, but she held her ground. Sebelius went to Washington in 2009 to become the secretary of health and human services, and, unfortunately, her successor brokered a deal to build a new 895 MW plant. Whether or not it will be built, however, remains to be seen. It may, in fact, become subject to regulation under the EPA's greenhouse gas regulations.[64]

Endangerment

The U.S. Environmental Protection Agency has taken a long road to regulating greenhouse gases. In 1998, during Bill Clinton's administration, the EPA first determined that carbon dioxide could be regulated under the existing authority of the landmark Clean Air Act. In 2003, under George W. Bush, EPA denied a petition asking it to regulate GHGs from new motor vehicles and then reversed its earlier view, from the Clinton administration, that regulating

GHGs falls under its authority. A coalition of states and environmental organizations sued the Bush EPA. In 2005, a panel of federal circuit court judges upheld the Bush administration ruling. The case was appealed before the Supreme Court, and that body said on April 2, 2007, that "greenhouse gases fit well within the Act's capacious definition of 'air pollutant,'" and that the EPA has a "pre-existing mandate to regulate 'any air pollutant' that may endanger the public welfare."[65] In other words, the Supreme Court ruled that the EPA had not only the authority but the *responsibility* to determine whether or not GHGs may cause harm.

Not surprisingly, the Bush administration rejected further action, but on December 7, 2009, President Obama's EPA chief, Lisa Jackson, signed two findings, the first of which was an "endangerment finding." The *Federal Register* says:

> The Administrator finds that six greenhouse gases* taken in combination endanger both the public health and the public welfare of current and future generations. The Administrator also finds that the combined emissions of these greenhouse gases from new motor vehicles and new motor vehicle engines contribute to the greenhouse gas air pollution that endangers public health and welfare under CAA section 202(a). These Findings are based on careful consideration of the full weight of scientific evidence and a thorough review of numerous public comments received on the Proposed Findings published April 24, 2009.[66]

The EPA had received more than 380,000 comments.

The U.S. Department of Transportation (DOT) and the EPA moved jointly to produce rules and regulations to create new standards that would not only require GHG reductions but also mandate improvements in fuel economy. It is very important to note that the new rules were substantially negotiated with the auto industry, including the United Auto Workers and elected officials from Michigan, the home of the American auto industry, as well as with the various states and environmental organizations that had been party to the

---

* The six greenhouse gases are carbon dioxide (CO_2), methane (CH_4), nitrous oxide (N_2O), hydrofluorocarbons (HFCs), perfluorocarbons (PFCs), and sulfur hexafluoride (SF_6). These are the same six gases that are regulated under the Kyoto Protocol.

lawsuit. Among the benefits of the program will be not only the reduction of greenhouse gases by nearly a billion tons of CO_2eq over the lifetime of the vehicles sold between 2012 and 2016, but savings of 1.8 billion barrels of oil as well. EPA put the monetary value of the program at $240 billion because of fuel savings and avoided health impacts, among other factors.[67] In August 2011, EPA and DOT followed up by announcing new standards for heavy-duty trucks and for buses. This sector accounts for almost 6 percent of all U.S. GHG emissions.[68]

But the Obama administration had to move to the next level: controlling industrial sources of GHGs. As we have seen, the U.S. Senate failed to advance climate and energy legislation in 2010. The subsequent Republican onslaught in the House of Representatives, and the diminishment of the Democratic majority in the Senate, was a death knell for substantive climate legislation in the 112th Congress that would run from 2011 through 2012—and very likely the outlook would not change for at least for the next several years. So, even though President Obama and his top administration officials like Lisa Jackson had repeatedly emphasized their preference for federal legislation to form the framework for ongoing control of GHGs, they were left with no option but to advance a regime via the regulatory route provided by the Clean Air Act—in fact, mandated by the law.

Not waiting for Congress, and perhaps anticipating the eventual outcome there, the EPA published its rules for the Greenhouse Gas Reporting Program in the fall of 2009. Twenty-five thousand tons of CO_2eq per year of the six regulated GHGs is the threshold for the requirement for an industrial facility or a power plant to submit an annual report. This program will identify and measure 85 percent to 90 percent of America's total greenhouse gas emissions, from thirteen thousand facilities in forty-one industrial categories. The deadline to submit 2010 reports via the electronic Greenhouse Gas Reporting Tool (e-GGRT) was September 30, 2011.[69] A comprehensive and well-vetted inventory is obviously a critical prerequisite for an effective greenhouse gas control regime.

In spite of the considerable opposition from politicians in the United States, mostly Republicans, ideologically motivated for the most part, but also many Democrats, principally those representing the special interests of oil, coal, and coal-dependent electric utilities, the EPA moved ahead on its regulations. Standards for fossil-fuel power plants and petroleum refineries were first up in the queue. These two sectors represent 40 percent of the GHGs in the United States. The new standards are scheduled to be issued in 2012 after extensive discussion with the industries involved and, given the volume of comment on the endangerment finding, a massive amount of public input on these regulations. Lisa Jackson had this perspective: "We are following through on our commitment to proceed in a measured and careful way to reduce GHG pollution that threatens the health and welfare of Americans, and contributes to climate change. These standards will help American companies attract private investment to the clean energy upgrades that make our companies more competitive and create good jobs here at home."[70]

Beyond the GHG standards, though, there are other regulations that EPA is enhancing or newly bringing into being that are having an impact on the power industry in particular. Under the authority of the Clean Air Act, Clean Water Act, and Resource Conservation and Recovery Act, these include a tightening of the existing standards for the acid rain precursor pollutants, sulfur dioxide (SO_2) and various compounds of nitrogen and oxygen (NO_x); new standards for mercury and other hazardous air pollutants; coal ash; and cooling water intake and power plant effluent.[71]

Clearly, the strengthening of environmental safeguards by the federal government, many of these decades in the offing, are driving a transition in the United States away from a dependence on coal for power.

Other Drivers of the Downward Trend for Big Coal

With or without EPA regulation of greenhouse gases, the overall trend of proposed coal plants being withdrawn from consideration or blocked has continued in the United States. In 2009, U.S. electricity generation fell by 4 percent, owing largely to the economic recession. However, over the same year, coal-fired power generation fell by 11 percent. There was at the same time a 7 percent jump in hydropower generation and 12 percent in renewables.[72]

Why? One big reason is that coal is being phased out, and new, high-efficiency natural-gas-powered power plants are taking over. Progress Energy, for instance, will shut down eleven of its coal plants in North Carolina— 1.5 GW worth of electricity—and replace them with gas. Other plants they operate have been retrofitted with emissions controls that have substantially reduced emissions of mercury, a potent neurotoxin, and nitrogen oxides and sulfur dioxide — the two principal precursor pollutants of acid rain.[73] Siemens is providing the gas units, and the head of its power generation group says that gas replacing coal is very much the trend.[74] Clearly, this is one company that is anticipating more extensive and intensive regulation of key air pollutants.

There appears to be a newly found abundance of natural gas contained in shale rock deposits in North America, Europe, China, and South America. The potential for Europe alone is a staggering 92 billion barrels of oil equivalent.[75] Legitimate concerns exist regarding the environmental and public health implications of getting this gas out of the shale in which it is embedded. However, if safely and cleanly produced, this gas can serve as a critical bridge to a decarbonized future, something that the environmental prophet Barry Commoner called for in his book *The Politics of Energy* in 1979. Gas, many energy and environmental analysts today agree, can be the solution to the immediate problem of the massive carbon dioxide burden from coal — not to mention the other environmental, public health, and economic nightmares that coal engenders. Natural gas also has much lower air pollution impacts than gasoline as a fuel for surface transportation and is, by and large, much cheaper to use. Recognizing the benefits, big commercial fleets like those of Walmart, FedEx, ups, and others are making the transition.

An MIT study confirmed the environmental bang for the buck inherent in a transition from coal to gas. Gas emits roughly half as much carbon dioxide as coal in a conventional central power plant.[76] With new turbines, that ratio improves, and with cogeneration, the value in reduced CO_2 output is even greater. The CO_2 intensity of natural gas plants has been decreasing for fifty years, especially so since the 1990s, while the same measure has increased for coal facilities. In fact, as the highly regarded energy analyst Sean Casten notes, "CO_2 emissions associated with a MWh of coal-derived electricity are 18 percent higher today than they were in 1960."[77] Given a level playing field — in other words, a "price on carbon" — the MIT researchers predict a virtual phasing out of coal use in the United States by 2035.

The Regional Greenhouse Gas Initiative in the Northeast United States, the Western Climate Initiative, and the Midwestern Greenhouse Gas Reduction Accord, involving a number of U.S. states and Canadian provinces, are all geared to reduce greenhouse gas emissions, the principal target being CO_2 from coal-fired power plants (see Chapter 5). In the process, coal and other high-carbon facilities are going to be retired.

U.S. Energy Secretary Steven Chu said early in 2011: "We're going to see massive retirements within the next five, eight years. Much of our fleet of coal plants is 40 to 50 years old."[78] Clearly, the downward trend in the United States for coal is not a product merely of the financial crisis and subsequent recession.

The Activist Factor

Another important factor in coal's loss of ground has been the work of activists. In the United States, the Sierra Club's "Beyond Coal" campaign calculates that over 150 plants have been taken off the drawing boards since 2001. It was at that time that Vice President Dick Cheney's energy task force, properly known as the National Energy Policy Development Group, recommended the construction of hundreds of new coal-fired power plants. The Sierra Club, working on a number of fronts, set out to defeat the proposed facilities, one by one, and with the indispensable participation of local activists. By early 2011, the abandoned plants represented just over 85 GW of power that would have produced 611 million tons of CO_2 annually.[79]

Mary Anne Hitt, director of the Sierra Club's Beyond Coal campaign, described how getting all these plants off the table happened: "We have done this by engaging in every possible venue — at the state level with public utility commissions, getting the story into the media, putting pressure on governors and other elected officials, to litigating when we thought that permits being issued violated the law. It's been a different set of factors that have won the day in each one of these campaigns. We've had an unprecedented level of success." The heart of the effort has been, Hitt emphasized, community organizing.[80]

The Sierra Club, the other major national environmental organizations involved in these fights, and the scores of local groups who have a stake in preserving their communities from the depredations of activities like mountaintop-removal coal mining and the dangers of air pollution from smokestacks and water pollution from coal ash lagoons, among other environmental and public health concerns, don't come to the table empty-handed when they are seeking to prevent new plant construction or, increasingly, the phasing out of old, dirty facilities. These activists bring with them a set of recommendations that encompass the entirety of the clean-tech portfolio of cost-effective, proven methods of saving and generating power. Hitt points out that in one state, Illinois, where 10 GW of proposed coal-fired power has been taken off the table, a nearly equal amount of wind is going to be brought on line. In fact, 12 GW is staged for development.[81]

Giving the Beyond Coal campaign an enormous boost was the four-year, $50 million commitment by Bloomberg Philanthropies in July of 2011. New York City mayor Michael Bloomberg's gift will triple the geographic scope of the campaign, from fifteen states to forty-five. At the announcement for the donation, he noted, "If we are going to get serious about reducing our carbon footprint in the United States, we have to get serious about coal. Ending coal power production is the right thing to do, because while it may seem to be an inexpensive energy source the impact on our environment and the impact on public health is significant."[82]

So, a whirlwind of forces has changed the momentum in the United States on electricity from coal: regulatory concerns, litigation, grassroots activism, the price of natural gas, and, not the least, financial risk. One top international banking executive had this observation: "Coal is a dead man walkin'. Banks won't finance them. Insurance companies won't insure them. The EPA is coming after them. . . . And the economics to make it clean don't work."[83]

Global Activism

The impulse to work diligently and well toward a clean energy future is not, by any stretch of the imagination, a uniquely American phenomenon. Activists all over the world have been involved in scores of nonviolent protests, legal actions, and focused organizing in the past several years. Most of the action has taken place at the grassroots level. What one key activist, Ted Nace, author of *Climate Hope: On the Front Lines of the Fight against Coal* and coordinator of the "CoalSwarm" Wiki, identifies in these protests is the creation of political space for policy makers to make the changes necessary to avert climate catastrophe.[84]

Another world-class activist, journalist and author Bill McKibben, founded 350.org to help spread the message about climate change and how to solve the crisis. The group has a global presence, and in the fall of 2010 organized a "global work party" in which there were over seven thousand events in 188 countries.[85] McKibben, along with the top leadership from Greenpeace USA and the Rainforest Action Network, issued a call for "creative, nonviolent protests and mass mobilizations."[86]

In the late summer of 2011, thousands of Canadian and American activists set in motion by 350.org descended on Washington and performed civil disobedience, resulting in nearly thirteen hundred arrests over a two-week period. In the fall, they were back and encircled the White House in a stunning visual demonstration of their commitment, calling on the president to deny the permit for the Keystone XL pipeline project, which would have greatly enhanced the commercial viability of the Alberta tar sands, thus enabling their further development. These nonviolent demonstrations, along with the fact that the U.S. Department of State received over 280,000 comments on its environmental impact statement by June of 2011[87] and editorial boards like those of the *New York Times* vociferously opposed the project, helped to clear the "political space" necessary to allow the bold decision that President Obama made on January 18, 2012. He denied the permit.

We saw an outpouring of protest focused on the climate crisis at Copenhagen in 2009, and before that in March of 2001 when George W. Bush announced the U.S. withdrawal from the process of entering into the Kyoto Protocol. European leaders expressed their outrage, as did their constituents by flooding the streets. The environment minister of Sweden said at the time, "No country has the right to declare Kyoto dead." France's environment minister called Bush's decision "completely provocative and irresponsible." The European Union Commission president remarked, "If one wants to be a world leader, one must know how to look after the entire earth and not only American industry." The head of Germany's largest environmental organization, Naturschutzbund Deutschland, said Bush's decision was "a catastrophe."[88] Bush's EPA administrator at the time, Christine Todd Whitman, said some years later: "The way it happened was the equivalent to flipping the bird, frankly, to the rest of the world on an issue about which they felt so deeply. It was something that just sort of said: 'Well, we really don't care. It's not important to us.'"[89]

In Europe, the activist record on fighting pollution is a particularly strong one. Philip Shabecoff, the first and foremost environmental journalist for many years, noted the importance of environmental politics in the unshackling of the chains that the Soviet Union maintained on its people and the people of Eastern Europe: "Environmental activists were the advance guard of the democratic revolution that changed the face of geopolitics in the late 1980s and early 1990s."[90]

The message, very simply, is never to underestimate the power of environmental activism.

Beyond the United States: Tightening the Noose on Coal

In Canada, a shift is well under way. Federal regulations that came into effect in 2011 will require the fifty-one coal plants in Canada, producing 19 percent of the electricity there, to shut down.[91] France will close down half its coal plants and double its output from renewables by 2020. RWE AG, one of Germany's top two electric utilities, won't be building any new coal-fired power plants, and Vattenfall, the largest utility in Scandinavia, will be converting its coal plants in Denmark for biomass.[92]

For Australia, where the impacts of climate change are a stark reality for many hundreds of thousands of its citizens dependent on farming and tourism, a carbon tax will be brought to bear to reduce emissions from all six Kyoto gases coming from utilities, transport, and industry. The coalition government is aiming for July 1, 2012, for beginning the program, and then transitioning within three to five years to a market-based mechanism. All the revenue will be devoted to managing the transition to a clean energy economy, offsetting the costs for families and businesses, and seeing to various adaptation programs to deal with the impacts that are in train.[93]

In the United States, one factor in the downward trajectory for coal plants is switching to natural gas facilities for power. There are other good approaches to fuel switching. In Britain, its biggest coal-fired power station, Drax, has been increasing its uptake of biomass. Drax generates 7 percent of the UK's electricity. Twelve percent of the fuel it has been using is now biomass, not coal. Biomass generates 90 percent less CO_2 than coal, surpassing even the 50 percent reduction one gets from using natural gas.[94]

You get, not incidentally, a lot more bang for the buck too when you use biomass to generate electricity and heat than when you try to convert it to biofuels for internal combustion engines, an as-yet difficult and costly process. As noted in Chapter 3, the internal combustion engine itself is highly inefficient, losing about 80 percent of the energy that goes in. Electric vehicles running on biomass-generated power would optimize the energy content of the resource.

We can see more key areas where the transition away from coal, however slow it may be in some cases, is taking place.

In India, the government is well aware of the burden of coal and the need for lowering its economy's carbon intensity. It began taxing coal, both mined in India and imported, in July of 2010 and expects to generate $650 million a year for its National Clean Energy Fund.[95] This is a significant step for a very important G20 nation.

It is, as virtually every economist who has studied the problem of climate change has asserted, essential to set a "price on carbon." Lord Nicholas Stern called climate change "the greatest market failure the world has seen," and the 2006 *Stern Review on the Economics of Climate Change* from the UK government said, in no uncertain terms: "First, we must establish a carbon price via tax, trade and regulation—without this price there is no incentive to decarbonise."[96]

For China, where we have seen that the carbon dioxide output from coal combustion has increased by 200 percent in the ten years from 2000, the stakes are high. The Chinese know this. They have, for instance, taken hundreds of old and inefficient power plants and factories out of service. This, along with the building of new clean-energy power plants, has served to reduce coal consumption by nearly two billion tons over the five-year period from 2006 through 2010.[97] That sounds pretty good, but the annual consumption in just one year, 2009, was 3.47 billion tons.[98]

The newest Chinese five-year plan, for 2011 through 2015, promises a continued emphasis on energy efficiency and renewables, even as the economy continues to accelerate like a bullet train. Among the new industries to which the Chinese made a commitment with this plan are clean energy and high-performance, high-MPG cars. China's second-highest official, Wen Jiabao, made a point of saying that the carbon intensity targets promised in Copenhagen had to be met, and that growth had to incorporate environmental protection. China's chief international climate negotiator said the new plan "will place climate change and green, low-carbon development at an even more important strategic position, raising countries' shared confidence in facing climate change."[99]

Nukes

The catastrophic events of March 2011 in Japan delivered a well-deserved shock to nuclear power worldwide and a long-overdue wake-up call to people everywhere. Nuclear power has been, since its inception in the 1950s, wasteful of financial resources and slow to build, has produced massive amounts of radioactive waste — which has never been adequately provided for in terms of its proper final, long-term disposal, and which, as we have learned once again, is inherently dangerous in the extreme.

Although the so-called "nuclear renaissance" of the years prior to Fukushima was often touted as a way to "decarbonize" energy production, it was far from that. For one thing, although there are certainly no greenhouse gas emissions from the day-to-day operations of a nuclear power plant, one must take into account the full "life-cycle" footprint. One exhaustive analysis of over a hundred life-cycle studies of nuclear plants from around the world, after averaging the estimates from the nineteen most-reliable studies, put the GHG output at 66 grams of carbon dioxide equivalent per kilowatt-hour (gCO_2eq/kWh). Although much lower than the 960 gCO_2eq/kWh from coal facilities or the 443 gCO_2eq/kWh from natural gas-fired plants, nuclear still fell well below the footprint of 32 for PV or 10 for onshore wind farms.[100]

Amory Lovins of the Rocky Mountain Institute has done extensive analyses of not only the utter lack of cost-effectiveness of nuclear power but its role in actually *slowing* our progress toward effectively dealing with the climate crisis at hand. In sum, Lovins asserts,

> expanding nuclear power is uneconomic, is unnecessary, is not undergoing the claimed renaissance in the global marketplace (because it fails the basic test of cost-effectiveness ever more robustly), and, most importantly, will reduce and retard climate protection. That's because — the empirical cost and installation data show — new nu-

clear power is so costly and slow that, based on empirical U.S. market data, it will save about 2–20 times less carbon per dollar, and about 20–40 times less carbon per year, than investing instead in the market winners — efficient use of electricity and "micropower," comprising distributed renewables (renewables with mass-produced units, i.e., those other than big hydro dams) and cogenerating electricity together with useful heat in factories and buildings.[101]

At a debate on nuclear power's efficacy in late 2010, Robert Alvarez, another leading expert on and outspoken opponent of this technology, had a number of compelling points to make regarding the monstrous costs involved in our global sixty-year experiment with nuclear power. Cleaning up, or even just stabilizing the wastes from nuclear power and weapons programs will require hundreds of billions, if not trillions of dollars. Building new nuclear power in the United States has a price tag of trillions. Beyond that, the loan guarantees that the U.S. government wants to extend for new plant construction? They have a greater than 50 percent chance of default. Alvarez said that nuclear power is a "millstone" holding back the flourishing of other, better technologies.[102]

In the wake of the Fukushima disaster, the costs of the clean-up and compensation for victims will number in the scores of billions of dollars. The Tokyo Electric Power Company (TEPCO) has asked the government for $12 billion to help with the initial compensation payouts as part of a much larger fund that TEPCO itself will help bankroll.[103] It will take at least thirty years and $20 billion to clean up the reactor facilities.[104] Prime Minister Yoshihiko Noda says that the Japanese government will spend at least $13 billion to clean up the areas around Fukushima that have been contaminated.[105] An analysis by U.S. and European experts a little less than a half year after the accident estimated that the facility may have released twice as much radiation as previously thought.[106] This finding obviously has implications for the long-term health impacts from radiation exposure and the ongoing displacement of more than eighty thousand people from their homes.

This catastrophic accident jolted the Japanese into an awareness of the dangers inherent in this technology. Seventy percent of the Japanese people now oppose nuclear power. The country is now fully intent on phasing out nuclear power and making a rapidly accelerated changeover to renewable energy. Other countries that have pulled the plug on nuclear power since Fukushima include Germany, Italy, Belgium, and Switzerland. In India, where plans on the drawing board were to increase nuclear generating capacity by ten times over the decade from 2010 to 2020, opposition has been building for several years, even prior to Fukushima. Grassroots opposition to new plants has been intensifying, as has concern from powerful regional political leaders. In October 2011, a petition was filed by leading citizens and civic organizations in India's Supreme Court to block new building at least until cost-benefit analyses have been done.[107]

Nuclear power may yet be fully recognized around the world as an unequivocally inappropriate antidote to greenhouse gas emissions from the power sector.

Where the Action Is

We have been looking a lot here at the big picture: policy being made at the national level by very big players such as the United States, China, India, and the EU. It is important to remember—and we have certainly touched on it numerous times—that an enormous amount of activity is taking place at the subnational level. Cities, states, provinces, and regional alliances and compacts of these political entities are changing the landscape on climate and energy. Some might call this the "bottom-up" approach.

One recent book, *Local Action*, highlights the many developments below the federal level of government in the United States. This movement has been growing exponentially, not only in the United States, but worldwide. As the authors of *Local Action* say: "This grassroots approach to emissions reduction is a powerful tool for combating climate change." Further, "We need to reduce emissions where they start—in cities—and it is up to local governments to take action."[108] That has been precisely what has been taking place.

We have seen, for instance, how California—which would have the ninth-largest economy in the world if it were its own country—is anchoring the initiative to build a robust cap-and-trade regime with itself at the core and a number of western states and Canadian provinces also involved. The world's carbon markets topped $120 billion in trades in 2010. Bringing

California and these others on board would make this a significantly more active market. The eleven participants in the Western Climate Initiative have a cumulative GDP of around $4 trillion, which would make them the fourth-largest economy.[109] California also took the initiative on regulating GHGs from automotive emissions, leading to the U.S. EPA and DOT negotiating a deal for the whole country.

As renowned scientist and activist Michael Oppenheimer noted in 2007, at the time of the release of the first part of the IPCC's Fourth Assessment Report, because Congress has not been able to effect federal legislation to curtail greenhouse gases, that didn't mean we were being hamstrung in our efforts. "Washington has been a political vacuum on this." But Oppenheimer was quick to point out some of the important developments in California, as well as the Regional Greenhouse Gas Initiative in the Northeast United States and the EU's work.[110] (Around the same time, Bill McKibben called Washington's sloth on climate and energy "twenty years of inactivity — a remarkably successful bipartisan effort to accomplish nothing."[111])

Actually, as we have been seeing throughout, there has been a tremendous amount of progress on engaging the issues and moving to deal with them. From finance to politics, from international negotiations — both multilateral and bilateral — to technological advances, from business and industry to the nonprofit sector, there has been a growing surge of activity. Not the least of this oceanic energy has been taking place in cities and states across the world.

The Golden State

California very often leads America, and indeed the world, in social, cultural, or political trends. Certainly on the question of environmental protection, particularly with air pollution, the Golden State has more often than not been out in front. Arnold Schwarzenegger, movie star and bodybuilder, may have been an unlikely candidate to play a starring role on climate and energy, but that is precisely what happened. *Time* magazine declared the Governator one of its "Heroes of the Environment" for 2008, saying "he's a global salesman for the war on carbon, spreading the message that you don't have to be a girly-man to help save the planet."[112] It's cute, of course, but Schwarzenegger earned it. That he has been one of the increasingly rare Republicans who has been progressive on climate and energy makes his views and accomplishments a little more noteworthy.

The "Million Solar Roof Initiative" is, for example, a hugely ambitious undertaking that is not only changing the face of energy generation and distribution in California, but globally. Because of the pure purchasing power of California's homeowners and real estate developers, the growth of solar there stimulates the industry everywhere. Growth has been phenomenal. In 2000, there was not quite 9 MW of grid-connected PV installed. By 2008, that number was 440.[113] The goal is to reach a million roofs by 2016. A "net metering" law came into effect in 2011 that will require utilities to buy surplus electricity from individuals. This will, as it has in Germany, Spain, and several other places, greatly spur the industry.

The solar roof initiative is part of a broader California Solar Initiative. There is also a solar heat and hot water program that offers rebates on systems for homeowners of nearly $2,000 for a system, and up to $500,000 for multifamily dwellings and commercial real estate. The California Solar Initiative seeks to install almost 2 GW of new solar by 2017. This is being underwritten by $2.17 billion from the state over the course of the program.[114] California also is shooting for 33 percent of its electricity to be sourced from renewables by 2020. The Renewable Portfolio Standard (RPS) is a strong incentive for utilities to build out their renewable infrastructure. As of 2010, the three big state investor-owned utilities together averaged 18 percent of their sales from renewables. In short, California has a vast, well-financed, and well-conceived program to enhance efficiency, conservation, and to produce low-carbon power.[115] With the election of Jerry Brown as governor in 2010 and Gavin Newsom as lieutenant governor, the stalwart environmentalism of the previous administration will not decline.

Newsom was the mayor of San Francisco, one of the world's greenest cities. San Francisco, like most of California's cities, is playing a big role. San Franciscans are reducing energy use, for instance, through building and lighting retrofits. They are actively studying how to use the enormous potential of wave and tidal power. They have made a commitment locally to reducing GHG emissions by 20 percent from 1990 levels by 2012. They have also set a particularly ambitious goal of a zero-emission mass transportation fleet by

2020. San Francisco recovers 77 percent of its waste through recycling, reuse, and composting, shooting for zero waste by 2020.[116] All these programs are cutting edge.

Jerry Brown served as the mayor of Oakland, San Francisco's sister city across the bay, for eight years. Oakland is perennially ranked in the top ten on any number of lists of green and sustainable cities. The Natural Resources Defense Council, for instance, named it a 2010 "Smarter City for Energy" because, among other things, it gets 17 percent of its energy from wind, solar, biomass, and geothermal, and because it wants to cut oil consumption in half by 2020 and also reach zero waste.[117]

Led by Mayor Antonio Villaraigosa, Los Angeles is not to be outdone in California. It has a goal of reducing GHGs 35 percent from 1990 levels by 2020. Owing to its topography, its industrial activity, and above all to its sprawl and reliance on automobiles, LA has had the worst air pollution of any city in the United States since the 1970s when the federal Clean Air Act came into being. The air quality has made quantum leaps since then, but Angelenos feel there is still room for improvement. Their attention to environmental quality has also predisposed the local population to take further strides toward sustainability. Because the LA metropolitan area is the thirteenth largest in the world and the second largest in America, what happens there matters. There are nearly thirteen million people there, four million of whom are in Los Angeles proper. The ports take in almost half of U.S. imports.

Their "ClimateLA" plan is, not surprisingly, ambitious. The municipal utility, the Los Angeles Department of Water and Power, will produce 35 percent of its power from renewable sources by 2020. LA had met its previous goal of 20 percent by 2010.[118] Green building and lighting retrofits are mandated under ClimateLA. Reducing water consumption by 20 percent is another important target for the largest city in the increasingly water-starved American West. New parks, a million new trees, and the revitalization of the Los Angeles River are all on the table. Greening the airports and seaports, with their enormous traffic of passengers and freight, is another priority.[119] LA, like the state of California and other cities there, is on a mission to heighten sustainability and reduce climate-forcing gases.

The Empire State and the Big Apple

California is not, by any stretch of the imagination, where sustainability begins and ends in America. New York State, for one, has a "Climate Action Plan" and the goal of reducing GHGs by 80 percent from 1990 levels by 2050. To do this, New York is working on a number of fronts. It is, for one thing, as we saw in Chapter 5, one of the charter members of the Regional Greenhouse Gas Initiative, or RGGI, that is lowering the cap, year by year, on emissions from electric utilities in the Northeast. Half a billion dollars from RGGI auctions will have gone to energy efficiency and clean-tech programs in the Empire State by the end of 2011. New York has an RPS of 30 percent by 2015, with funding to help stimulate scores of large renewable projects. New York also has an energy efficiency standard of 15 percent — to reduce the projected demand by that amount by 2015, with $4 billion to be spent to achieve that. New York also recognizes the inevitability of some changes in the natural and built environments as a consequence of climate change and has, among other things, a sea-level-rise task force in place to bring proper focus to bear.[120]

In the Big Apple, America's largest city, Mayor Michael Bloomberg's "PlaNYC" was rolled out on Earth Day in 2007. It has been rolling along, with an emphasis on green building and transportation. On the green building side, new laws and regulations have been put into place to require significant work on energy audits and retrofits. There is a vibrant green building movement among architects and developers in New York City. As evidence, you need only look, as we did in Chapter 3, at the Empire State Building retrofit that will reduce energy use by 40 percent and the new Bank of America Tower, one of the world's greenest buildings.

Mass transportation is integral to New York City's economic well-being. New York City's buses and subways rank number one in the country for passenger miles traveled, with a whopping 11.9 billion annually. NJ Transit, serving NYC and urban and suburban New Jersey, is number two, clocking 3.5 billion passenger miles traveled. Third? The Metro-North Railroad system with trains in and out of Grand Central up to the northern suburbs and exurbs and into Connecticut. Miles? Some 2.18 billion. (Fourth place goes to the Metro system in Washington, DC, with 2.1 billion passenger miles traveled in 2008.)[121] New York State has the lowest per capita energy consumption of any of the fifty U.S. states. There is a simple reason why: New York City's extraordinary public transportation infrastructure.

Even with this extensive system in place, the Metropolitan Transportation Authority (MTA) is continuing to expand and upgrade. Of the 8.5 million average weekday ridership on the MTA's vast system of subways, buses, and trains, 7.4 million is on the NYC subways.[122] One of the biggest and most important initiatives is to build another subway line on the densely populated East Side of Manhattan. The Second Avenue subway has been talked about for decades, particularly since the demolition of the Second Avenue elevated line — the "el" — and the Third Avenue el after that. This new line, when it is finally built in the next decade, will ease congestion for hundreds of thousands of riders and help take thousands of buses and cars off the roads.

New York City needs to improve its performance on surface transportation if it wants to continue to be a world leader in commerce, finance, and tourism. It also needs to continue to reduce its air pollution burden. Pollutants like ozone and carbon monoxide are hugely diminished since the 1960s, to be sure, as is the case with almost all developed-nation cities, but there is still room for improvement in local air quality. Beyond that, in order to meet its goal of reducing GHGs 30 percent from 2005 levels by 2030, NYC wants to reduce its emissions from transportation by 44 percent by that time. It has introduced a truly robust program to create bicycle lanes, and it also wants to bring electric vehicles into the mix. To that end, the Mayor's Office of Long-Term Planning and Sustainability is actively exploring how to create charging stations and bring EVs to New York's streets.[123]

Congestion Pricing

Mayor Bloomberg failed, however, in one of his most ambitious undertakings: introducing a congestion pricing program to reduce traffic in the Big Apple. One estimate has put the annual price tag on excessive traffic in New York City at $13 billion, owing to delays, wasted fuel, lost revenue, and the increased cost of doing business.[124] In order to mitigate that problem, in any urban environment, one tack to take is to levy a fee for the privilege of driving in a city's central business district during weekday business hours. Bloomberg's plan, after having been recommended by the City Council, with the support of scores of municipal good government, environmental, labor, and business groups, and with hundreds of millions in federal money available to support the program, sailed up the Hudson River from New York to Albany, the state capital, where it died. (To make a long story short, the reason for the demise of the Big Apple's congestion pricing plan was that the second-most-powerful elected official in the state, Assembly Speaker Sheldon Silver, killed it. He was getting pressure from his members from the four boroughs of the city outside Manhattan and from the suburbs — plus Silver has an abiding, intense dislike for Bloomberg. Thus, too often, the shape of New York politics and public policy.)

In a number of other major metropolises, London, Stockholm, and Singapore among them, congestion pricing has thrived, doing what it was intended to do and gaining the support of the vast majority of citizens. In London, more passengers are using public transport than before the plan went into effect, and £148 million ($240 million) was raised as revenue in the city's 2009–10 fiscal year, to be devoted wholly to improving the mass transit system. Two-thirds of Londoners polled approve of the charge. In Stockholm, a high-tech system reads license plates and charges vehicles accordingly. The city has experienced a 20 percent drop in traffic and a 10 percent decrease in air pollution. Singapore has seen a 13 percent reduction in vehicles and a 22 percent rise in average traffic speed.[125]

Asian Cities

Singapore, not incidentally, has been judged Asia's "greenest city." Using an array of criteria in categories including energy and CO_2, land use and buildings, transport, waste, water, sanitation, air quality, and environmental governance, Singapore was the only one of twenty-two cities, from Karachi to Tokyo, and from Beijing to Jakarta, that scored well above average. Not surprisingly, rich cities did better than the poorer ones in the "Asian Green City Index" commissioned by Siemens and formulated by the Economist Intelligence Unit. As with so many enterprises of this pith and moment, a number of key collaborators have helped design the criteria by which the cities have been graded. In this case the OECD, the World Bank, and CITYNET, Asia's regional network of local authorities, were involved.[126]

"The Asian Green City Index supports cities in their efforts to expand their infrastructures on a sustainable basis. We want to enable Asia's up-and-coming urban centers to achieve healthy growth rates coupled with a high quality of life," said the chief sustainability officer for Siemens.[127] This is the quintessence of sustainable development: enhancing quality of life, including protecting public health and the natural environment, coupled with economic growth.

Beyond this, one of the key findings of the study was that environmental awareness has been growing steadily, most of these cities have an array of environmental regulations in place, and that they are being enforced. "Cities that performed well in the Index are characterized by their ability to successfully implement environmental projects and consistently enforce regulations," said the lead researcher for the study.[128]

Siemens has invested a tremendous amount of intellectual capital in creating tools for boosting the sustainability of urban centers around the world. It has published a study on "Megacity Challenges," interviewing five hundred top city managers and planners in twenty-five selected urban environments, looking at trends and challenges as well as citing best practices; produced a series on sustainable urban infrastructures, zeroing in on how to use resources and abate CO_2 efficiently; and is continuing in its series of Green City indices. The Asia study was preceded by Europe and Latin America, with Germany, the United States, and Africa rolling out in 2011.[129]

South of the Border: Latin American Cities

The Latin American Green City Index looked at seventeen cities. Brazil's Curitiba, with 1.7 million people, known since the 1960s for its far-seeing vision of sustainability, ranked number one among the cities studied. Planning, for Curitiba as for the other cities that made high marks, is the key. Curitiba, for one, integrated initiatives on solid waste management and economic development, as well as air pollution control and public transportation. Interestingly, unlike Asia, good environmental performance did not necessarily correlate with high per capita GDP. Again, planning seems to be a critical factor in setting the more successful cities apart.[130]

Mexico City, with a population over nineteen million, is hoping to effect a major transition to sustainability in its "Plan Verde." Mexico City is committed to spending a billion dollars (U.S.) a year through the next dozen years in order to develop a number of programs on air, water, waste, and transport, among others, and to radically reduce GHG emissions. Mass transit is a critical component, with further expansion of subways, electric streetcars, and the very effective and increasingly popular Bus Rapid Transit (BRT) system.[131]

A Big Ticket: Bus Rapid Transit

BRT has been proliferating all over the world as a consequence of the fact that it is much cheaper to build and operate than heavy rail, subways, or light rail (streetcars). BRT systems often achieve the same speed and volume, with the comparable comfort, of these heavier, more traditional systems. BRT travel is quicker than regular bus travel because the buses use dedicated lanes and the passengers get on and off more quickly.[132] BRT systems are well entrenched or are being planned on all six continents, with 120 cities up and running, and with many more than these under construction or being planned. In existence now are 280 corridors along 4,300 kilometers (2,672 miles), served by 6,700 stations and 30,000 buses, and serving about twenty-eight million passengers per day. Major systems exist today in Bogotá and Curitiba, Beijing, Jakarta, Brisbane, and Rouen, to name but a few leading examples. Shanghai, Tehran, Glasgow, Bologna, Buenos Aires, Rio de Janeiro, New York, Chicago, and Seattle are among the scores of cities with systems being planned or under construction.[133]

BRT has enormous potential for reducing greenhouse gases and is an approved methodology under the Kyoto Protocol's Clean Development Mechanism, making BRT also a big potential moneymaker for cities. Bogotá's program, for instance, receives nearly a quarter of a million Certified Emission Reduction (CER) credits each year under the CDM.[134] These CERs are, of course, fungible on the carbon markets.

European Cities

As we have seen, Europe has been in the vanguard in the fight to address the climate crisis. The EU, its member states, and other European nations have been pushing the world to act on energy and climate, and have taken the bull by the horns at home. This leadership is manifested in many instances in what Europe's cities have been advancing.

Siemens and the Economist Intelligence Unit have also taken a comprehensive look at sustainability in Europe's cities. They looked at thirty cities in thirty countries, using much of the same criteria as they have for the other studies on Latin America and Asia. Scandinavia did particularly well, with Copenhagen, Stockholm, and Oslo as the top three. As with the other studies, awareness was a key indicator of how a city performs. The local populations where the environment and public health are a primary concern do better. They support sustainability initiatives with their pocketbooks, at the voting booths, and in their lifestyles.

The study is clear about why cities matter. Half the people on our planet now live in cities, and the cities are responsible, through food, energy, and water consumption, transportation, construction, and other factors, for as much as 80 percent of our GHGs. Most of the cities studied have a target for GHG reduction. Copenhagen has the startling goal of being zero carbon by 2025. Again, as with the other studies, wealth is a good predictor of how the locality performs overall.[135]

Networks and Compacts

It is beyond the scope of what we are able to look at here to provide in-depth analyses of the hundreds of exciting initiatives and programs taking place all over the world in cities, provinces, and states. From transportation and energy, water and sanitation, green open space and habitat, these subnational jurisdictions are leading the way toward sustainability in myriad ways.

Much of the work of supporting these efforts internationally, financially and through the sharing of best practices, has been taken up by nongovernmental organizations like ICLEI — Local Governments for Sustainability, an association of over 1,220 local governments, from seventy countries, representing nearly six hundred million people. (ICLEI was founded in 1990 and originally named the International Council for Local Environmental Initiatives.) Through conferences, research projects, training courses, the publication of case studies and other reports, and with the ability to provide technical support from its expert staff, ICLEI provides local governments and other interested stakeholders with tools to further sustainable development. From Argentina to Zimbabwe, from small rural towns to major world cities, ICLEI promotes local action on climate.[136]

Another highly effective player in this arena is the Climate Group, which has a raft of corporate members and state and provincial governments who are engaged on projects ranging from promoting EVs and green IT to monitoring the progress of the Climate Principles, guidelines adopted by several major global financial firms.[137] Similarly, the Institute for Local Self-Reliance, in business since 1974, helps communities in the United States on waste management and distributed generation of energy, among other areas.[138]

The World Bank, not surprisingly, is very engaged on this agenda: to transform cities into places where people and the planet are working in continuous harmony. Its report *Cities and Climate Change: An Urgent Agenda* notes that the world's fifty most-populous cities have a combined GDP of almost $10 trillion, second only to America's $14.7 trillion. These cities are directly responsible for 2.6 billion tons of CO_2eq annually.[139] The good news, though, is that large cities are efficient in terms of their per capita emissions of GHGs. The trick, therefore, is to optimize our resource efficiency globally by continuing the trend of environmental sustainability evidenced by the hundreds of examples that the more visionary and aware cities are providing for us.

The C40 Cities Climate Leadership Group, working in close partnership with the Clinton Climate Initiative (CCI), has been yet another effective vehicle for promoting best practices and giving material support to these better and smarter ways of doing the business of our cities. Summits of leaders from these largest of the world's cities have been taking place every other year since 2005, with the meeting for 2011 taking place in São Paulo. Bill Clinton's group, the CCI, has been providing considerable technical, project, and purchasing assistance along with financial advice.[140] One partnership, for example, has brought together cities, banks, and energy services firms to retrofit buildings for maximum efficiency. Launched at the C40 Summit in New York City in 2007, this program has boosted over two hundred projects with over half a billion square feet of building space in nearly fifty cities.[141] (That's 175 Empire State Buildings worth of space.)

The City Climate Catalogue is an inventory of targets for GHG reduction from almost three thousand communities globally, big cities and small, and the actions being taken to meet those targets. It is heartening because it illustrates, yet again, the broad sweep of the communities committed to effecting movement toward sustainability. It shows that diverse peoples from all over the world recognize what is at stake and are ready, willing, and able to make the changes necessary to alter the "business-as-usual" course we have been on for so many years.[142]

The U.S. Conference of Mayors (USCM) was founded during the height of the Depression to help direct relief to American cities desperately in need. It represents today over twelve hundred cities with populations over thirty thousand and works on a broad array of fronts including social welfare, economic development, and tourism, energy, and the environment. In 2005, Seattle mayor Greg Nickels initiated the USCM's Climate Protection Agreement in which cities agreed to undertake actions including committing to meet GHG-reduction goals and lobbying their state and federal elected representatives to create new programs, including a market-based national emissions reduction program. They also agreed to put a new organization into place to help produce their desired ends: the USCM Climate Protection Center. In 2005, 141 mayors signed the compact, and the list has grown to 1,049 today. Not the least of the Climate Protection Center's achievements has been helping to shape the Energy Efficiency and Conservation Block Grant Program. Under the economic stimulus package enacted in 2009, $2.8 billion will be used for this program, profiting hundreds of American cities.[143]

New Perspectives

It is certainly true that there is an attitude of outright antipathy toward environmental regulation in the United States evidenced by most of the leadership and the rank and file of one of its two major political parties, namely the Republicans, also known as the Grand Old Party. This is particularly manifest in the GOP's hostility toward regimes to control greenhouse gases. The official Republican position opposes cap and trade.

There are, of course, a few Republican leaders who have supported smart policies on climate and energy, Arnold Schwarzenegger and Michael Bloomberg among them. But, except for one Republican presidential candidate, Jon Huntsman, all the rest of the field in the run-up to the 2012 elections either denied the existence of the climate crisis or have waffled. Mitt Romney is the poster child for the wafflers: "Do I think the world's getting hotter? Yeah, I don't know that, but I think that it is. I don't know if it's mostly caused by humans."[144] The views of these candidates have been driven by the virulently denialist Tea Party movement. The Republicans are, for the most part, the only major political party on Planet Earth that contests the reality of anthropogenically induced climate change.

This party is indeed isolated and alone. William Hague, the UK foreign secretary, a Conservative, has said: "Climate change is one of the gravest threats to our security and prosperity. Unless we take robust and timely action to deal with it, no country will be immune to its effects." Angela Merkel, Germany's chancellor, a Conservative (with a PhD in physics), has said: "We cannot afford failure with regard to achieving the climate protection objectives scientists tell us are crucial. That would not only be irresponsible from an ecological point of view, but would also be technologically short-sighted, for the development of new technologies in the energy sector offers major opportunities for growth and jobs in the future." Nicolas Sarkozy, president of France until 2012, a Conservative, said: "To sum up, we have to choose between disaster or the solution. We are deciding for the whole planet and what we don't decide, those who follow us will no longer be able to decide. Rarely has a choice been so crucial for the future of mankind." These views are shared by virtually all the top leadership in government in the world today, as well as by mayors and governors from all the four corners of the earth and everywhere in between. It is the express opinion of all the major multilateral organizations, from the UN to NATO, from the World Bank to the OECD, that climate change is a threat to peace and development and that it must be met head-on.

Further, this perspective is shared by governments and their leadership, high and low, in the developing world as well as in the developed. We have seen, for example, that the Chinese and the Indians are fully engaged in developing alternatives to business as usual.

There has been a revolution in consciousness. We are aware as never before of the dangers of climate change and the opportunities in transitioning to smarter, cleaner, safer, healthier, less costly, and more secure ways of doing business. We are, thankfully, and in spite of the reactionary politics of an increasingly isolated minority in the world, making substantial progress toward true sustainable development.

a lighter footprint

CLIMATE CHANGE AND SUSTAINABLE DEVELOPMENT

The War on the Forests

We have seen the devastation visited on the Amazon rainforest over the past couple of decades. Photographs and film have shown us the indelible impact of our folly. This stark picture of deforestation on an epic scale has been reproduced in images from the Indonesian and Malaysian peatlands, and the forests of the Congo Basin and West Africa as well. We have also witnessed the havoc played on the boreal forests of Alberta from tar sands extraction and the destruction of hundreds of thousands of acres of mature Appalachian hardwood forests by mountaintop-removal mining.

There are four billion hectares of forest in the world, covering 31 percent of the land. That is 15.4 million square miles. The five largest countries in the world also account for more than half of the total forested area.* Forest losses from economic activity and from natural causes averaged 16 million hectares (61,800 square miles) per year in the 1990s, an area slightly more than Nepal and slightly less than Tunisia. From 2000 through 2010, the loss was around 13 million hectares (50,200 square miles). Most of the forest loss has been taking place in tropical regions.[1]

Aside from the destruction of habitat and loss of biodiversity, the depletion of resources on which millions of indigenous peoples rely, the impacts from soil erosion and resultant flooding, and the pollution from fires, deforestation has been generating a great portion of the greenhouse gases that have been steadily forcing the climate system. On average, emissions caused

---

* Russia, Canada, the United States, China, and Brazil, with over twenty-one million square miles in total.

by deforestation during the 1990s were estimated at about six billion tons a year of carbon dioxide.[2]

Deforestation is primarily driven by agriculture, but also by opening up territory for settlement and road building, and by unsustainable logging.[3] In Brazil, there has been enormous pressure on the Amazon rain forest from cattle ranching and soybean cultivation. For Indonesia, and also Malaysia, wetlands dominated by peat, hugely biodiverse and massive carbon sinks, are being drained and burned to make way for palm oil plantations. Rainforest is also being decimated in these Asian countries. One way to look at this is that if you count land-use changes—not just emissions from industrial, energy, transportation, and agricultural activities—Indonesia would trail only China and the United States as a contributor to greenhouse gases. Brazil is fourth in the world.[4]

One of the greatest threats in deforestation is, of course, the loss of biodiversity. We are, in many ecologists' estimation, in the Sixth Great Extinction on earth. One distinguished scientist, Dr. Norman Myers, asks the question, "How much time do we have left before the mass extinction underway surpasses our best efforts to contain it?"[5] Note that he does not equivocate about the fact that there is a "mass extinction underway." Myers identifies tropical forests as particularly key to the issue, as they contain easily more than half of the world's total of species and that the destruction of these forests is happening at a greater rate than any other ecosystem.

A study published in *Nature* in March 2011 confirms this fear: overall, the current rate of extinctions is occurring at a pace well above normal. The lead scientist for the study, paleoecologist Dr. Anthony Barnosky, author of the book *Heatstroke: Nature in an Age of Global Warming*, noted the principal cause: "The current rate and magnitude of climate change are faster and more severe than many species have experienced in their evolutionary history."[6]

How do you stop a runaway train? That question that Myers asks, in the context of biodiversity, is also relevant when you are talking about what is happening in the Amazon relative to drought. Since the 1960s, as much as—or more than—20 percent of the Amazon has been razed.[7] Because of the protracted and extensive deforestation and degradation there, rainfall itself has been affected. Rainforests are themselves net producers of rain. The

Amazon forest makes half its own rain and 40 percent of the rain in the rest of Brazil. As world-renowned ecologist Thomas Lovejoy said: "It has been obvious from the beginning that deforestation could at some point cause this hydrological cycle to unravel."[8]

Agriculture and hydroelectric plants in the region rely on this water. But Amazon rain is also important beyond Brazil, affecting the world's climate system as a whole. The Amazon experienced devastating and unusual droughts in 2005 and 2010. When this happened, the ability of the rainforest to absorb the billion and a half tons of CO_2 that it normally does in a year was greatly circumscribed. The trees that died stop serving as carbon sinks, they rotted and further generated carbon dioxide, and much of the remaining vegetation was diminished in its capacity to breathe in carbon dioxide in the photosynthetic process. It is entirely possible that the Amazon could "flip" from being a net consumer of atmospheric CO_2 to a net producer, with obviously dire implications for the climate system as a whole.[9]

Because the drought kills vegetation, it exacerbates warming locally and globally, at the same time reducing the overall moisture of the ecosystem, leading inevitably to more drought, and so on. This vicious cycle is known as a "positive feedback."[10] These feedback loops exist in other key climate-system contexts, such as with the melting of Arctic sea ice, the increase in water vapor — the most prevalent greenhouse gas — in the atmosphere because of higher temperatures, and the thawing of permafrost. This cycle is not "positive," of course, in the sense that it is producing a desirable outcome, but merely in the sense that it is self-reinforcing.

The Beetles

Another key positive feedback that we have been seeing in our forests is taking place in the North American West. From British Columbia to New Mexico, tens of thousands of square miles of forest have been decimated by beetle infestations. In British Columbia alone, where the insects have hit the hardest, mountain pine beetles have affected over sixty-three thousand square miles.[11] Why has this unprecedented level of attack by beetles been in evidence over the past decades? The extensive coniferous forests of the West have been subjected to less rainfall and warmer temperatures. The drought conditions make the trees weaker and more susceptible to these predatory species, and the warmer air, particularly during winter, leads to much lower mortality for these beetles. Here is how the scientists say it: "Because bark beetle population survival and growth are highly sensitive to thermal conditions, and water stress can influence host-tree vigor, outbreaks have been correlated with shifts in temperature and precipitation."[12]

This is another vicious cycle in which the beetles prey on weaker stands of trees, further increasing their own population and enabling their spread. As the trees die and rot or burn, they release their carbon back to the climate system, they are no longer able to sequester carbon, and they are no longer able to hold moisture in their structures and in the surrounding soil.

This is not unimportant, because American forests, as elsewhere, have an essential function in sequestering carbon. In the United States, trees consume as much as 19 percent of the carbon dioxide emissions from fossil fuel combustion.[13] (Numbers vary on this between forestry experts, but the range is from 12 to 20 percent.) As we saw in Chapter 2, land-use changes, including deforestation, accounted for the preponderance of the anthropogenically induced greenhouse gases introduced into the climate system from the time of the Industrial Revolution until the early part of the twentieth century. That changed as formerly agricultural land was reclaimed by forests as urbanization arose. This trend is reflected in figure 7.1, a graphic from the U.S. Forest Service illustrating how carbon sequestration by American forests has fluctuated over time.

That vital function could well shift back to a net emission from American forests if stressors like drought, wildfires, and beetles continue to ravage them, particularly in the West.

The Tar Sands

Aside from the beetles threatening to move east from British Columbia to Alberta, that latter Canadian province has another pernicious problem of deforestation in the boreal zone with the development of bitumen tar sands by oil companies.

The northern, or boreal, forests of East Asia, Russia, Scandinavia, the United States, and Canada are the largest terrestrial ecosystem in the world.

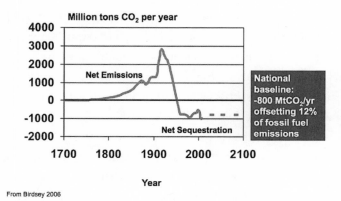

7.1 Carbon budget of U.S. forests (forest ecosystems and wood products). Dr. Richard A. Birdsey, Forest Service of the U.S. Department of Agriculture

They comprise an area equal to 10 percent of the world's land surface. As other forests do, the boreal forests capture CO_2 in photosynthesis and store it in the living vegetation and soil. These lands, though, have accumulated carbon over thousands of years and stored it in peatlands and wetlands, in permafrost — land frozen for more than two years — and even in ice in deposits known as methane clathrates. The carbon stored in the northern forests and wetlands is nearly twice that stored in tropical rain forests and about six times that stored in the temperate forests.[14]

Under a full-development scenario in the tar sands regions of Alberta, natural areas including forests, peatlands, and wetlands being disturbed would amount to 1.6 million hectares, or nearly 6,200 square miles. These natural areas would be given over to the mining operations, plus the waste ponds, processing facilities, roads, and other infrastructure. If all the carbon stored in this area were released due to deforestation and other forms of land degradation, it would translate to 2.12 billion tons of CO_2. A conservative estimate, however, puts the total release at closer to 873 million tons.[15]

Tar sands oil is, not incidentally, much more greenhouse gas intensive in its extraction and processing than conventional oil. Canada is producing over 1.3 million barrels of oil a day, the majority of which is being exported to the United States. The developers hope to nearly triple that amount to 3.5 million. If three million barrels of tar sands oil replaces the equivalent amount of conventional oil in the mix, that would be tantamount to adding twenty million new cars to our highways.[16] (Of course, when all the life-cycle impacts are added in, the carbon dioxide burst from burning all the developable tar sands oil for fuel is the biggest fraction.)

Mountaintop-Removal Coal Mining

In the mountains of rural Appalachia in the eastern United States, over five hundred separate surface coal mining operations have destroyed more than a million acres. What happens is simple: Coal companies plant explosive charges on the tops of pristine mountains in Kentucky, West Virginia, Virginia, and Tennessee and then level them to extract the coal from seams that lie below. They push the spoils into the adjacent valleys, thus ruining the water sources for local residents. If the forest cover continues to be blown up in this fashion, one estimate puts the loss of biosequestration capacity at over three million tons of CO_2eq a year.[17]

Yet that is a drop in the bucket compared with the loss of the hugely productive carbon sinks that exist in the world's coastal wetlands and mangrove forests. One study calculates the loss of mangroves to be 30 to 50 percent over the past half century. Given the capacity these ecosystems have to store carbon, much more than the boreal, temperate, and tropical forests at which we have looked, that may translate to as much as 120 million tons of carbon given off to the climate system annually—nearly 440 million tons of carbon dioxide.[18]

Making Peace with the Forests

Runaway train? You bet. The world, however, has been moving with greater focus and ever more swiftly to curtail deforestation and other land-use changes that are exacerbating climate change. There are many more ways of addressing these problems today than there were twenty years ago when the world met in Rio for the Earth Summit. We have been slow, to be sure, in getting at the problems, but the rates of deforestation and other degradation have been decelerating, and there is considerable reason to believe that we can reverse our momentum toward catastrophe.

Had we gotten an earlier start on this, of course, it would have been better for everyone. The Earth Summit in Rio de Janeiro in 1992 produced the UN

Framework Convention on Climate Change. What it did not generate was a Convention on Forestry. The United States, among a number of other countries, supported a broad agreement to curtail deforestation. What did come out of Rio was a much weaker "Non-Legally Binding Authoritative Statement of Principles for a Global Consensus on the Management, Conservation and Sustainable Development of all Types of Forests."[19]

William K. Reilly, the head of the U.S. delegation to Rio, wrote at the time that "developing countries are simply not yet ready for it, fearing a threat to their sovereignty in the global concern for better forest conservation and sustainable use. I was frankly startled by the depth of developing countries' anxiety about the industrialized world's concern for forests. Many poorer forest-owning nations genuinely fear an 'internationalization' of their natural resources."[20]

Reilly said, twenty years later, that a convention "could have been a unifying thing." The politics and the costs for a comprehensive regime to lower greenhouse gases, as we have seen since Rio, are particularly daunting for the United States, as they are for a good number of other nations. Reilly has felt all along that the politics of a forest agreement "would be seen as more congenial to the United States. The purpose, which would have clear, direct wildlife values, would be more compelling politically to Americans, and, of course, the cost could be more constrained, more limited. I always thought that was a missed opportunity. But there it is. We're still working at it."[21]

In Kyoto in 1997, the protocol, as with the original convention, did not allow for a program to prevent deforestation. This approach to climate change mitigation was conspicuous by its absence from the Clean Development Mechanism. Many concerned NGOs wanted to see precisely this way of addressing climate change fully included in the CDM. Reilly recalls: "It was something that was supported by World Wildlife Fund." (Reilly was chair of World Wildlife Fund USA at the time.) "We couldn't swing the international community."[22] The CDM, though, as we will see, was finally empowered in 2001 to accept reforestation and afforestation projects.*

---

* According to the IPCC's AR4 glossary, *reforestation* is the planting of forests on lands that have previously contained forests but that have been converted to some other use. *Afforestation* is the planting of new forests on lands that historically have not contained forests.

A solid working plurality of the world community's policy makers, finally, has seen the wisdom and value of slowing, stopping, and reversing deforestation, not only for curtailing the 20 percent of the total of manmade global greenhouse gases that result from land-use changes, but for preserving biodiversity, saving regional ecosystems from desertification, and generating employment and tremendous income from sustainable forest management, agroforestry, and ecotourism, among other co-benefits.*

REDD

The 194 parties to the UN Framework Convention on Climate Change have been slowly but steadily moving to incorporate forestry programs into their work. At the Bali meetings in December 2007, avoided deforestation was embraced as a necessary focus going forward. The overall approach has been deemed Reducing Emissions from Deforestation and Forest Degradation—or REDD. In September of 2008, the UN launched its REDD program under the aegis of the UN's Food and Agriculture Organization (FAO), the United Nations Development Program (UNDP), and the United Nations Environment Program (UNEP), with a secretariat for the interagency compact in Geneva.[23]

The program had collected nearly $100 million in its trust fund as of 2011, of which a little more than half had been transferred to participating organizations for their projects. The Norwegians have been, by far, the biggest donors to this fund, having deposited more than $84 million. The three UN agencies leading REDD have received the funds disbursed to date and directed them for capacity building with the UN system and to big forest nations such as the Democratic Republic of Congo, Indonesia, Vietnam, and Tanzania, among others. Participants are engaged in a host of activities, including alliance building and dialogue, creating technical capacity and planning REDD strategies, designing the critical components for monitoring and assess-

---

* *Co-benefits*, also according to the IPCC's AR4 glossary, are the benefits of policies implemented for various reasons at the same time, acknowledging that most policies designed to address greenhouse gas mitigation have other, often at least equally important, rationales (e.g., related to objectives of development, sustainability, and equity).

ment at the country level, and developing methodologies and tools, as well as information-sharing modes at the UN level.[24]

We are in the first phase, then, of a three-phase approach to fully implementing REDD. This first phase is known as the "readiness phase," in which countries are creating strategies for their forest programs and building expertise to eventually implement them. The process will culminate in the full deployment of these programs with the flow of funds to finance them.[25]

REDD+ and Beyond

Prior to and at the Copenhagen meetings in 2009, the parties decided to go beyond REDD by incorporating more than just avoided deforestation and forest degradation in also seeking to address the conservation of forest carbon stocks, sustainable management of forests, and enhancement of forest carbon stocks. This is known as REDD+. There are now twenty-nine developing countries in the Asia-Pacific area, Africa, and Latin America eligible to receive REDD+ assistance. These countries are receiving technical and financial support for their national programs from various developed nations and international funds. REDD+ is, as yet, in its infancy, but it is growing. It is projected that as much as $30 billion a year may be flowing to forestry projects in the not-too-distant future in order to realize the multiple benefits involved.[26]

The French and Norwegian governments, in the spring of 2010 after Copenhagen, took the initiative. In March of that year, France hosted the International Conference on the Major Forest Basins, and in May, Norway held the Oslo Climate and Forest Conference. From these was launched the REDD+ Partnership, seventy-one developed and developing nations that are either supporting or undertaking significant forestry programs. The work of the Partnership substantially complements the work of the UN agencies.[27]

REDD+ is but one framework, though, for any number of financing initiatives that are trying to substantially address the difficult problem of deforestation and other land-use changes that are the cause of billions of tons annually of greenhouse gas emissions as well as the destruction of vital habit for wildlife and livelihoods for people.

The Clean Development Mechanism, for instance, has been slowly building up its capacity over the past several years to foster reforestation and afforestation projects. As of November 1, 2011, there were fifty-nine afforestation and reforestation projects under way, plus three projects seeking to protect mangroves. These projects are eligible to receive almost five million certified emission reduction (CER) credits a year, ostensibly offsetting the same amount of greenhouse gases measured in tons of CO_2 equivalent. There are twenty approved methodologies, with dozens more under consideration.[28]

The CDM is still coming to maturity on forest projects. Meanwhile, there are other pathways to alleviating the stresses that we have been putting on our forests. Multilateral development banks (MLDBS) are very much involved. The World Bank has, for example, its BioCarbon Fund (BioCF), which leverages mostly private capital, along with public money from both foreign governments and those where projects are being developed, and money from NGOS, in order to support a range of reforestation, afforestation, and avoided deforestation initiatives. The BioCF supports twenty-five projects in five regions of the world.[29]

For example, one of the World Bank–financed projects, in partnership with World Vision, a Christian humanitarian organization around since 1950, is the "Humbo Assisted Natural Regeneration Project" in Ethiopia. Begun in 2007, it is expected to capture almost nine hundred thousand tons of CO_2eq over thirty years. It is, like so many mitigation and adaptation initiatives taking place under the auspices of the CDM or in other sustainable development contexts around the developing world, a profoundly community-oriented program. It employs "farmer-managed natural regeneration" techniques to restore forest, at the same time stimulating considerable economic development and protecting streams.[30]

The World Bank is working in partnership with other MLDBS such as the African Development Bank, the Asian Development Bank, the European Bank for Reconstruction and Development, and the Inter-American Bank, to roll out projects through the Climate Investment Funds. One of these is the Forest Investment Program, which has earmarked nearly $600 million for programs in Africa, Latin America, and Asia.[31] Another vehicle is the Forest Carbon Partnership Facility (FCPF), which has fifteen donor nations and one organization, the Nature Conservancy, that have pledged $345 million for, among other things, "REDD+ Readiness." This includes projects such as building capacity among indigenous peoples to be involved with forest carbon activities. The FCPF will also devote money directly to initiatives that are able to make verified emission reductions.[32]

MRV

The idea of MRV — the measurability, reliability, and verifiability of carbon reductions — is critical. How do you measure the effectiveness of preserving a particular forest area? What is the biosequestration capacity of the types of vegetation in the area under review? Over the course of how much time will the forest be protected and sequestering carbon? How do you know that the forest is, in fact, being protected? MRV is the vehicle through which you address these key questions. Aid money and project development financing through the Clean Development Mechanism and other market-based approaches cannot be properly calibrated and disbursed without full faith in the process.

The IPCC, for one, established guidelines in 2006.[33] These and other methodologies and tools for MRV show that, as one paper puts it, "Scientists, governments and NGOs have made substantial progress towards addressing technical issues surrounding REDD, including how to ensure that REDD is real, verifiable, permanent and benefits regions of intact and non-intact forests."[34] One way of doing this is with remote sensing via satellite. Key organizations involved in this work include the UN's FAO, NASA, the European Space Agency, and the U.S. Geological Survey. Remote sensing combined with the inventorying of resources from the ground can, with the technologies available today, provide highly accurate pictures.[35] The Global Terrestrial Observing System (GTOS) is one enterprise at the heart of these efforts. GTOS, a collaborative of UN agencies such as FAO, UNEP, the World Meteorological Organization, and others, observes, models, and analyzes "terrestrial ecosystems to support sustainable development."[36]

An exciting and important new tool for measuring forest carbon is lidar,* a radar-like technology that uses lasers deployed from a plane. One tropical ecologist, Gregory Asner of the Carnegie Institution, says, "It's kind of

---

* Lidar is an acronym for light detection and ranging.

like an MRI. We can figure out not just the height of the canopy, but also the 3-D structure of the forest." With satellites, on-the-ground inspections, and lidar, scientists can get highly accurate assessments of the carbon in the forest biomass.[37]

The Palm Oil Factor

As we have noted, Indonesia's rainforest and peatlands destruction, plus the forest fires that are routine there, qualifies that country as the third-largest contributor of greenhouse gases. Most of this environmental havoc occurs as a consequence of the creation and expansion of palm plantations. Palm oil is used for food products such as margarine and for cooking oil, and for detergents, cosmetics, and, more and more, biofuel. The world production of palm oil is nearing fifty million tons a year, about half of which comes from Indonesia, with another 37 percent coming from Malaysia. Indonesia exports 76 percent of its production and Malaysia, 91 percent. India, China, and Europe are their biggest markets, while the Indonesians and Malaysians themselves consume quite a lot.[38]

Given the extraordinary commercial value of palm oil — Indonesia alone exported over $14.5 billion in palm oil products in 2008 — the industry is expected to continue to grow. On the present track, production is expected to expand worldwide to sixty million tons by 2020.[39]

There have been, of course, many voices decrying the destruction of the forests and wetlands in Indonesia and Malaysia, not the least of the concerns being the tremendous pressures on the climate system, but also the debilitating air pollution that comes from the forest fires set to clear land for palm plantations, as well as the pressures on wildlife. The war for sustainable palm oil production and the preservation and, in some cases, restoration of these valuable lands is being fought on several fronts.

The Norwegians made a landmark commitment in 2010, pledging $1 billion for REDD programs for Indonesia. This is one more example of a vehicle for funding for these critical programs that complements the work of the UN and the multilateral development banks. Speaking at the Oslo Climate and Forest Conference in the spring of 2010, Indonesian president Yudhoyono said, "Working with our developed country partners, we will protect Indonesia's globally significant carbon- and biodiversity-rich tropical rainforests while helping local populations become more prosperous."[40]

Backing up these words, President Yudhoyono announced a two-year moratorium on clearing of land. However — and not surprisingly — the palm oil interests and other industries, such as mining, pushed back. What type of forest and other lands are at issue? Should the ban extend beyond the two-year period? Should lands already permitted for clearance be pulled back? There have been competing views within the government itself. After some initial delays, the program began to take shape.[41] A year after Yudhoyono's announcement, the government passed the law necessary for implementation, barring new permits for peatlands and primary forests. This protects 158 million acres — nearly 250,000 square miles.[42] From Norway's perspective, things are going very well. The Norwegian minister of environment and international development, speaking at an international conference on Indonesia's forests, said: "The president has issued an overall policy about how Indonesia will combat climate change. . . . What he has done today is a very positive step in making Indonesia a world leader in the fight against climate change."[43]

The commitment by Norway and others to help Indonesia preserve these critically important biodiverse and carbon-rich lands is an important step forward. The focus here is very sharp. At the same time, though, knowing that palm oil production, as well as the forest products and mining industries in these sensitive lands, will not soon be going into eclipse, and fully recognizing the economic value of such lands for these developing countries, various much more sustainable approaches to cultivation, logging, and mining are being brought into play.

The Roundtable on Sustainable Palm Oil (RSPO) is one such approach. WWF, formerly known as the World Wildlife Fund, got the ball rolling in 2001. Stakeholders held various preparatory meetings, and RSPO was formally established in 2004. There are now 441 full members, with participation from banks and investors, retailers, as well as environmental and development NGOs, and with the majority of members coming from the consumer goods manufacturing sector and the growers, processors, and traders of palm oil, from forty-one countries. RSPO has been fostering the use of sustainable practices all along the supply chain, from growing to processing to consumption. RSPO certifies the products as sustainable according to a series of guidelines developed by the stakeholders involved. More than five billion tons of palm oil and palm kernels, from twenty-four growers and ninety-nine mills, almost all of which are in Indonesia and Malaysia, have been certified. In addition, 135 facilities, being operated by seventy-four companies, have received supply-chain certification.[44]

There is, as we can see, great emphasis on producing palm products so as to minimize environmental impact. One influential organization, the Rainforest Alliance, while supporting the work of RSPO, has promulgated even tougher standards for palm products based on those of the Sustainable Agriculture Network (SAN). SAN was founded in the 1990s by several Latin American conservation and development organizations along with the Rainforest Alliance.[45]

Pressure on the palm industry has been particularly fierce coming from Europe. As part of their efforts to fight climate change, the twenty-seven members of the European Union are required to come to a 10 percent minimum for biofuel in their transportation sector. The EU has promulgated standards for biofuels, defining what fuels can be used to meet these targets. The standards explicitly bar the use of biodiesel made from palm oil that comes from plantations that have displaced rainforest or peatlands. The European Commission is clear in delineating that only biofuels that have a much lower GHG footprint relative to petrol or diesel will be eligible for public support. This does not equate to a big proportion of European biofuel—only about 5 percent—but it sends an important message.[46]

Most palm oil, in Europe and elsewhere, is used for food and industrial purposes, such as for cosmetics. Pressure from NGOs on the EU and on manufacturers has had an impact. The Malaysians, for example, report that their sales to Europe have gone down as a result of the activists' influence on the public.[47] Organizations like WWF flag companies on their use of sustainably produced palm oil and encourage consumers to buy from them. They have influenced major manufacturers to source their palm products from sustainable producers. United Biscuits, one of Europe's largest makers of cookies, crackers, and the like, is on board. It has been working with its suppliers and expects to have 100 percent certified segregated sustainable palm oil in 2012.[48]

Unilever, one of the world's biggest consumer goods companies, with revenue in 2010 of $55 billion, has recognized the problems associated with palm products, not only for the deforested and degraded lands of Indonesia and Malaysia, but for its own corporate profile—and sales. Unilever uses about 3 percent of the world's annual production of palm oil. Like other companies, it has committed to sourcing its palm oil from sustainable sources, and is a founding member of RSPO.[49]

But beyond that, Unilever is hoping to shift at least some of its reliance on palm oil to algae. In partnership with Solazyme, it is assiduously testing whether or not this is feasible within the foreseeable future.[50]

The activist impulse does not confine itself to high-powered international nongovernmental organizations. It exists in the hearts and minds of two Girl Scouts from Michigan. Rhiannon Tomtishen and Madison Vorva found that some Girl Scout cookies had a lot of uncertified palm oil content. Working with the Union of Concerned Scientists, the girls have gotten the Scouts highly sensitized on this matter, and Kellogg's, which makes the cookies, has at this point agreed to buy "green palm" certificates for 100 percent of its palm oil use, while it works to directly purchase RSPO-certified palm products. While they are at it, the two young activists are also working to get the Girl Scouts to mandate the use of more forest-friendly oils, like canola or olive.[51]

On a business-as-usual track, Indonesia could boost its already dangerous carbon footprint by as much as 60 percent in the next twenty years. However, given the spotlight that has been brought to bear on its land-use practices, and the growing willingness of Indonesia and other countries like Brazil, along with their partners in the developed economies, to confront these issues, the capacity for radically altering the picture is increasing all the time. Indonesia could, according to a government report, lower its GHG output by 70 percent within the same time frame. Three-quarters of those reductions could come from the land sector.[52] Given the tools that are becoming more available, such as the CDM, various REDD+ programs, and the flows of money from concerned and involved countries like Norway, Indonesia could be providing a massive GHG abatement, not only with economic benefits for its citizens, but with enormous positive implications for the climate system.

America's Contribution

Like Norway, the United States has been involved as a donor and as a vigorous proponent of sustainable forestry projects for the developing world. The Tropical Forest Conservation Act (TFCA) of 1998 created a program for grants to support an array of conservation activities. The TFCA is itself modeled on the Enterprise for the Americas Initiative from 1991. The Agency for International Development (USAID), along with some other agencies, administers the TFCA. The program will provide $239 million over the next twenty years via "debt-for-nature" agreements in fourteen countries in which debt owed to the United States is forgiven in return for specific conservation projects. USAID also supports programs as diverse as the Asia Regional Biodiversity Conservation Program in Vietnam, ecotourism in Uganda, and REDD capacity building in the Congo Basin.[53]

Pushing to Sustainability and Brazil's Progress

As difficult and complex as deforestation and land degradation in Brazil are for all the stakeholders involved, particularly seeing how much is at stake for the commercial interests there, primarily the cattle ranchers and agribusinesses growing soy for export, Brazil has made excellent progress in recent years on curbing deforestation.

The destruction and degradation of Brazil's Amazon rainforest, where the bulk of the deforestation for beef cattle has taken place, is, as we have seen, a tremendous driver of climate change. Although direct deforestation hasn't taken place in Amazonia as a result of Brazil's two other major agricultural commodities, soy for livestock feed and sugarcane for biofuel production, the pressure on land use from these crops has further exacerbated the decimation of the rainforest. Cattle ranching accounts directly for about two-thirds of Amazon deforestation, with small-scale farming the cause for much of the rest.[54] Brazil is the world's largest exporter of ethanol (from cane sugar),[55] as well as beef and soy.[56]

In spite of these pressures, data from 2009–10, including that from satellite images, show that annual deforestation diminished by 67 percent from the average of the decade from 1996 through 2005. The average annual forest loss went from 19,508 square kilometers (7,532 square miles) to 6,451 square kilometers (2,491 square miles). An analysis by the Union of Concerned Scientists equates the savings in CO_2eq to one billion tons.[57]

How did they do it? Pressure for government protection of the rainforest has come from a broad array of Brazilian groups, including those representing indigenous tribes, small rubber interests, and human rights and environmental organizations.[58] By December 2007, the pressure had grown sufficiently that Brazil proposed at the annual UN climate change summit the creation of an international fund to foster rainforest preservation in the Amazon. The Amazon Fund came into being in August 2008. Brazilian president Lula da Silva said then, "Brazil is conscious of what the Amazon means to the world." Lula da Silva sought $21 billion through 2020 for projects to reverse the trend on deforestation.[59] As of the fall of 2011, there were twenty projects contracted for or approved by the Amazon Fund, with $139 million earmarked in support. Much of the work is for capacity building by local governments and NGOs.[60]

In September of 2008, following on Brazil's creation of the Amazon Fund, Norway, the staunchest supporter of REDD+ by far among the world's nations, pledged $1 billion for Brazil's efforts through 2015. This, of course, galvanized the Brazilians to get down to work right away to make the fund operational. As a recent analysis of Norway's International Climate and Forest Initiative (NICFI) has it:

> NICFI's support has been effective in that it has successfully stimulated Brazilian environmental and climate policy debates and efforts to reduce deforestation. The fact that the Amazon Fund is widely regarded as an important example of the development of a national mechanism for disbursement of results-based payments, and [that] NICFI's support had a positive impact on momentum and direction of change in Brazil must also be recognised as successes.[61]

The Norwegians made a first payment of $110 million in 2009. In exchange for its payments, Norway expects to see measurable, reliable, and verifiable progress on slowing deforestation. That has been the case so far. The progress has been slow, to be sure, in identifying projects, but momentum is definitely building.

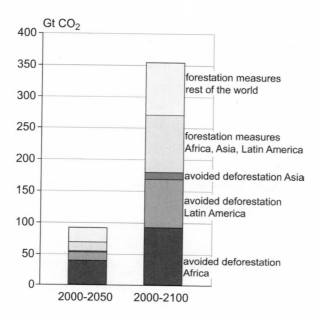

7.2 Cumulative mitigation potential, 2000–2050 and 2000–2100. "Climate Change 2007: Mitigation of Climate Change," Working Group III contribution to the Fourth Assessment Report of the Intergovernmental Panel on Climate Change, figure 9.6; Cambridge University Press

Beyond the critical area of funding is the role of NGOs like IPAM and IMAZON. They have done key monitoring and have helped educate the local agricultural and forest products workers how to maximize productivity while minimizing forest degradation. There has also been a dramatic expansion of areas protected from development by state and federal authorities, with effective enforcement. The Green Party has also been useful in helping to produce a positive direction. The Zero Deforestation campaign, founded in 2008, is seeking an end to deforestation in Brazil by 2015.[62]

The Sky's the Limit

Given a vibrant REDD+ program, the potential for not only reversing the trends in deforestation but actually using forests for the ongoing biosequestration of carbon dioxide is truly enormous. Figure 7.2, a chart from the IPCC's Fourth Assessment Report in 2007, is an excellent illustration of this.[63] It shows the mitigation potential in preventing deforestation and in fostering new forests. As you can see, the sky's the limit.

If forest destruction and degradation account for a fifth of the problem of global forcing of the climate, then there is also, by any measure, a huge upside for using forests to cool the planet. One of the key corollaries to this proposition is that while you are mitigating climate change, you are also improving the lives and livelihoods of the millions of people who depend on forests. Sustainable development, as we have noted a number of times, is a key aspect of how the world approaches its confrontation with the climate crisis. It is embedded in the message of the UN Framework Convention on Climate Change, as well as in the work of other key partnerships such as the Major Economies Forum on Energy and Climate. The Declaration of the Leaders of the Major Economies Forum on Energy and Climate in July 2009 explicitly stated that "moving to a low-carbon economy is an opportunity to promote continued economic growth and sustainable development" and, in fact, "low-carbon development is indispensable to sustainable development."[64] Sustainability is also a first principle in adaptation to the impacts of climate change. A joint statement from the science academies of thirteen of the world's leading nations said: "A strategic approach to adaptation must be based on the principle of sustainable development."[65]

The multiple benefits of REDD+ programs to address deforestation and forest degradation include, then, virtues that are sometimes characterized as "ecosystem-based." That is to say that in protecting the forests, you are also protecting and enhancing the biodiversity of life that exists within the forests, and you are preserving microclimates. You are at the same time preventing soil erosion and regulating the proper flow of water throughout the watershed of which the forests are an integral part, and allowing for the sustainable harvesting of forest products, including foods and other non-timber products. That generally translates into preserving the culture of indigenous people and improving their access, as well as access for others, to jobs, land tenure, and payments via various vehicles for preserving and protecting the forests.[66]

The Magic of Certification

The Rainforest Alliance has been a leader for twenty-five years in the fight to protect forests. It was the first organization to certify forest products as being sustainably produced. It helped found the Forest Stewardship Council (FSC), which now maintains the gold standard in certification guidelines, the only one supported by all major environmental groups. FSC has issued over a thousand certifications covering 141 million hectares in eighty-one different countries.[67] (That's 544 thousand square miles, an area larger than Peru but smaller than Mongolia.)

The Rainforest Alliance has taken a holistic approach all along. Its work to preserve forests and farmlands recognizes the critical component of ensuring the livelihoods of the people living in these areas. This is, of course, at the heart of sustainable development. While ecosystems are being protected, the Rainforest Alliance's certification standards also ensure that workers have safe and healthful conditions and receive fair wages, while smallholders receive fair prices for their goods. The Rainforest Alliance also helps tie these sustainable business operations into the growing global network of consumers seeking sustainable goods and services.[68]

Tensie Whelan, executive director of the Rainforest Alliance, agrees with William Reilly's assessment that the Earth Summit was a "watershed event in environmental history.

"It focused people on sustainable development, particularly civil society and the private sector, in a way that they hadn't before then. The real changes made since have been made through partnerships and engagement between civil society and the private sector."[69]

In many cases, the level of protection for forests and the sustainable payback for local people are greater in certified areas than in areas that have been protected by government mandate. For example, the Guatemalans created the Maya Biosphere Reserve in 1990. The area of rainforest that is under certification in the reserve, where various uses are allowed, but under the strict FSC standards administered by the Rainforest Alliance, has seen twenty times less deforestation and far fewer wildfires than even the areas under government protection where harvesting is prohibited.[70]

Whelan says, "Communities have a self interest in protecting the areas in which they have concessions. They have training and support in sustainable forest management, they have the market linkage so they can put food on the table. That's the kind of paradigm shift on which we're focused."[71]

Organizations like the Rainforest Alliance work with local NGOs to "train the trainers" in sustainable forest management. The Forest Stewardship Council certification standards that they administer, along with two dozen other organizations around the world, provide the matrix of how to do business so that local communities can both protect their resources and prosper. This paradigm shift that Whelan describes has been bringing new life to forest communities as the gospel spreads.

Here is how the FSC describes it: "Because managing forests the FSC way means following the highest social and environmental criteria, it often requires managers to adapt their management and operations. This is how FSC has a direct and permanent positive impact on the world's forests and the people living from, in and around the forest."[72]

Companies serving consumers, and consumers themselves, are also being increasingly better educated on what constitutes sustainability in the forest and farm products that they sell and use.

Many Ways to Skin a Cat

As we have seen, there is a vast and growing array of programs addressing forests. Jeff Hayward, climate initiative manager for the Rainforest Alliance, admits there is a lot going on in the universe of REDD+. "Creating the coherence, whether it's institutional arrangements, financial arrangements, the difference between developing country and developed country interests, or whether it's the tension between mitigation of climate change versus adaptation, all across the board, there's a great deal that's not going to fit together in a tidy compartment or box. It's probably better that it doesn't because it does need to be flexible and adaptive."[73] This echoes, as noted in Chapter 5, what George Soros said, in the context of the carbon markets, regarding the "bottom-up" approach to climate change mitigation that is likely to produce a multiplicity of prices for carbon emissions: "because there is a multiplicity of sectors and methods each of which produces a different cost curve."

The UN and other international agencies fully understand the need to maximize the economic productivity of the world's forests while at the same time protecting them. The FAO maintains that there are 13.7 million full-time jobs associated with the forestry sector in its three traditional subsectors (round-wood production, wood processing, pulp and paper) with $468 billion in gross value added worldwide.[74] These numbers do not reflect the value of crops that are grown in and around forests such as coffee, bananas, and cocoa, among scores of others, nor the value of the support businesses that grow up around the forestry sector, nor, for that matter, the "ecosystem services" of preventing soil erosion, maintaining biodiversity, and preventing degradation of water resources.

As global consciousness about the value of forests has continued to rise, local communities, governments, and NGOs have been seeing more and more success in slowing and in some cases stopping deforestation and even expanding forest cover. Several parts of the world have shown notable success. Europe's forests have grown over the past twenty years owing to two factors that are at work in other areas where success has been evident: new planting and the natural expansion of forests into former agricultural lands. Spain has been the best performer in Europe. Although Africa continues to deforest, the rate of reduction has slowed. Ten countries in Africa actually added forest in the twenty years between 1990 and 2010, with Tunisia, Côte d'Ivoire, and Rwanda topping the list. In Asia, as we have seen, deforestation and degradation have been particularly egregious in Indonesia and Malaysia where palm oil is being produced. Some areas of Asia, however, have made significant gains. China has been the leader because of its recognition of the critical importance of its forests for protecting watersheds, controlling soil erosion, and keeping the deserts at bay. Vietnam has been another excellent actor, having restored forest cover to forty-three percent of its land by 2010.[75]

One of the most important trends is the growth of "secondary" forests where farms existed previously. This was certainly the trend in North America in the late nineteenth and early twentieth centuries after the land that had been cleared in the East by the early settlers went back to forest because the heart of American and Canadian farming had moved west to the prairies, and the percentage of the urbanized population grew (see figure 7.1 earlier in this chapter).

Secondary forest growth is a particularly important factor in the tropics where deforestation has been rampant. One estimate is that nearly 2 percent of the world's tropical forests are in the process of regrowth.[76] A key paper on the subject puts it this way: "Current human demographic trends, including slowing population growth and intense urbanization, give reason to hope that deforestation will slow, natural forest regeneration through secondary succession will accelerate, and the widely anticipated mass extinction of tropical forest species will be avoided."[77]

Another critical positive trend is the reforestation and afforestation of enormous and growing swaths of land. From central government mandates like those from China to the work of schoolchildren from all around the planet, we have been planting trees at a dizzying pace in the recent past. One outspoken and enormously successful proponent of these programs was Dr. Wangari Maathai.

In 1976, Maathai, a biologist, started the Green Belt Movement (GBM) in Kenya. Since then her organization has inspired the planting of over thirty million trees there. GBM has more than six hundred community groups that operate more than six thousand nurseries. Her work has expanded well beyond Kenya to much of the rest of Africa. Maathai recognized the value not only of the trees themselves for restoring degraded lands, helping to preserve biodiversity and other environmental benefits, but also the importance for poverty reduction and community empowerment.[78] She won the Nobel Peace Prize in 2004 for her Homeric efforts.

Her campaign has also focused on creating new conditions in which women have been able to take the lead on building sustainability in their communities. Beyond this, the Green Belt Movement and Maathai helped build the democracy movement in Kenya that culminated in 2002 in the fall of the Moi dictatorship. Like the democracy movement in Eastern Europe in the 1980s, the environmentalists' message was a powerful part of the new politics in Kenya.

The chairman of the Norwegian Nobel Committee said at the award ceremony: "Maathai stands at the front of the fight to promote ecologically viable social, economic and cultural development in Kenya and in Africa. She has taken a holistic approach to sustainable development that embraces democracy, human rights and women's rights in particular."[79]

Maathai also inspired the UN's "Billion Trees Campaign," launched at the annual climate change conference held in 2006 in Nairobi, in which the UN Environment Program undertook to plant a billion trees as a way to involve people in better understanding the imperatives of climate change and the importance of fighting deforestation and forest degradation. The campaign quickly surpassed its original goal and had rocketed past twelve billion trees by the fall of 2011, with no end in sight. Millions of people have participated. Among the countries that have led the way in devoting themselves to the campaign are India, with 2.1 billion trees planted, and China, with 2.9 billion.[80]

The Chinese have been making the planet a lot happier by planting trees by the billions in recent years. From the time of Mao's infamous "Great Leap Forward" in 1958, the Chinese deforested relentlessly for nearly twenty years. Unfortunately, with the opening of China to the West in the 1970s and the advent of Deng Xiaoping's economic reforms in the late '70s, the pace of deforestation accelerated. This culminated in devastating floods in the Yellow and Yangtze river basins by the late 1990s. These areas lost 30 percent of their forest cover from the 1980s. Realizing how damaging the unrestricted logging had been, China reversed course and reassigned over a million workers into non-forestry industries, as well as devoting money and focus to reforesting and protecting remaining forest areas.[81]

The Chinese recognized before the 1990s that tree planting in the North would help prevent the continuing encroachment of the desert. Begun in 1978, the "Great Green Wall" is meant to cover twenty-eight hundred miles, buffering the cities, towns, and agricultural lands from desertification and helping to reduce soil erosion. Beyond this, the Chinese are hoping to restore the nation's forests so that they cover 42 percent of the land by 2050, an area slightly smaller than the combined size of India and Chile.[82]

There are also negative offsets in play, which, if they don't entirely cancel out the progress made, certainly serve to diminish the overall benefit. These include the fact that China's continuing insatiable demand for resources necessitates the import of enormous amounts of timber from its neighbor, Russia, leading to the deforestation of much of the Russian Far East looming in the near future. Much of the tree planting in China, it must also be noted, is in plantations. These do not conduce to the conditions for biodiversity that natural forests provide.[83] China also contributes to deforestation in tropical rainforests. It is, for instance, purchasing vast tracts of land in heart of the Congo Basin — 6.9 million acres (roughly the size of Haiti) — for conversion to palm plantations.[84] It is also the second-biggest importer of palm products in the world after India.[85]

Still, tree planting — since 1981 a requirement of citizenship for all Chinese between the ages of eleven and sixty — is unquestionably a tremendous boon to the country in terms of helping to reverse desertification, restoring degraded lands, fostering an environmental ethic, and even, to a certain extent, boosting ecotourism. Some scientists, both within China and elsewhere, assert that the scale of the reforestation, afforestation, and avoided deforestation there is having a palpable positive impact on countering the burden of greenhouse gases that China generates as a consequence of its epic, and growing, industrial output and consumption.[86]

Overall, increasing awareness, plus more money, focus, and expertise are being brought to bear. The UN declared 2011 the International Year of Forests. We are certainly not yet out of the woods (as it were), but as we have seen, there is great promise in REDD+ and a range of other programs being brought forward by international governmental and nongovernmental organizations, aid agencies, national governments, the private sector, and last, but most definitely not least, local communities.

As Dr. Maathai said in a recent documentary film: "Clothe the earth — put on the skin, a dress. A green dress, like trees, like vegetation. And then, when the earth is covered with green, with vegetation, it looks very beautiful. And in this age of climate change, can you imagine how happy the planet would be?"[87]

Food and Fiber's Impact

As we have noted, most deforestation in the modern world is driven by agriculture, both subsistence agriculture and industrial agriculture that exists primarily for export. But as we have also seen, as attention is brought to the issue, forestry practices can and do change. This potential for positive change is true for farming as well. We are seeing growing awareness of and concern for both what we consume and how we produce it.

The FAO published a landmark report in 2006 titled *Livestock's Long*

Shadow: Environmental Issues and Options. Some startling conclusions were made, among them the fact that livestock accounts for 18 percent of humankind's annual production of greenhouse gases. This 18 percent is attributable to a number of factors, including land-use changes such as deforestation for cattle and the massive land-use changes in rainforest and savanna for soy cultivation, grown for livestock feed. Grazing land and farmland to grow feed for livestock occupy very large percentages of the earth's terrestrial surface. There is, with growing world population and accompanying higher levels of affluence in many places, growing demand for meat, dairy, and eggs. Pressure on land also places pressure on biodiversity. In 40 percent of the world's ecoregions identified by wwf, livestock are listed as a "current threat." There is also considerable stress on water resources as a consequence of irrigation and grazing.[88]

What are the components of this startling pressure on the climate system? One calculation puts the burden of GHGs from deforestation from agriculture and livestock at nearly six billion tons of CO_2 annually. The incomplete uptake of inorganic fertilizers and manure lead to the production of nitrous oxide, accounting for 2.1 billion tons of CO_2eq. Methane from ruminants — cattle, goats, sheep, and some other common livestock — is as much as 1.8 billion tons of CO_2eq a year. Manure that is not applied to the land accounts for another 400 million tons of CO_2eq a year from both methane and nitrous oxide. There are other inputs of greenhouse gases from agriculture that are not necessarily related to livestock, such as the 600 million tons of CO_2eq a year in methane from rice cultivation and the hundreds of millions of tons in various other activities such as generating energy for the manufacturing of agricultural chemicals, for irrigation, and for working the land.[89]

The amount of livestock is staggering. Just in terms of animals killed for food worldwide in 2009, the numbers amounted to 1.7 million camels, 24 million water buffalo, 292 million cows, 398 million goats, 518 million sheep, 633 million turkeys, 1.1 billion rabbits, 1.3 billion pigs, 2.6 billion ducks, and 52 billion chickens. There are, of course, many billions more of many of these animals providing milk, eggs, and fiber.[90] Livestock accounts for 20 percent of the total biomass of all nonhuman terrestrial animals.[91]

Knowing the impact of livestock production on the climate system is a very important factor in then knowing how to better deal with confronting the crisis in which we find ourselves. Unfortunately, expert, high-level analysts of our climate situation like Lord Nicholas Stern, Rajendra Pachauri, and Al Gore have been pilloried for suggesting that one solution is to eat less meat. Pachauri, chairman of the Intergovernmental Panel on Climate Change, co-winner of the Nobel Peace Prize in 2007 along with Gore, has said: "In terms of immediacy of action and the feasibility of bringing about reductions in a short period of time, it clearly is the most attractive opportunity. Give up meat for one day [a week] initially, and decrease it from there."[92] Pachauri got e-mails saying his vegetarian diet left him malnourished, therefore he couldn't be thinking straight and thus was coming out with bad ideas.[93] That was among the milder reactions to Pachauri, Gore, and Stern's comments regarding meat consumption.

Nevertheless, the highly regarded PBL Netherlands Environmental Assessment Agency substantially buttressed Pachauri's statement with a paper in the prestigious peer-reviewed journal *Climatic Change*. The paper, "Climate Benefits of Changing Diet," asserted that the global transition to a low-meat diet would not only lessen the considerable greenhouse gas burden from manure and enteric fermentation, but would also radically reduce the GHG impact from land-use changes and "reduce the mitigation costs to achieve a 450 ppm CO_2-eq. stabilisation target by about 50% in 2050 compared to the reference case."[94] That means cutting the cost of saving the climate system by half with a simple lifestyle change.

Unfortunately, although the rate of meat consumption in the developed world appears to be leveling off, if not yet declining, in the developing world the pace is accelerating.

Biosequestration: Agriculture as Carbon Sink

Whether or not levels of meat consumption diminish overall remains an important question. However, there are scores of other ways to reduce the greenhouse gas footprint of our food and fiber production, many of which are coming into full flower.

The GHG mitigation potential in many of these practices is appreciated by agronomists, climate scientists, and activists alike. The potential, along a number of routes, is in the billions of tons of CO_2eq reductions a year (see figure 7.3 graph, from the IPCC's Fourth Assessment Report).

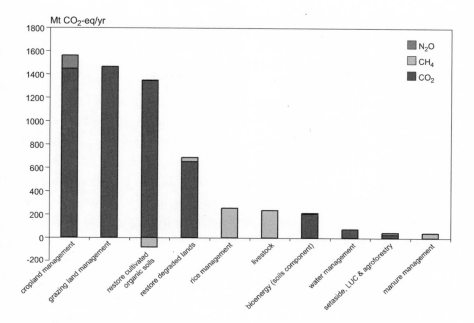

7.3 Global technical mitigation potential by 2030 of each agricultural management practice showing the impacts of each practice on each GHG. "Climate Change 2007: Mitigation of Climate Change," Working Group III contribution to the Fourth Assessment Report of the Intergovernmental Panel on Climate Change, figure 8.4; Cambridge University Press

Take, for instance, the category of cropland management. Within this category are several headings. Under the heading of agronomy are improved crop varieties, rotating crops, and using cover crops when land is not being used for active cultivation of food or other farm products.[95] Under nutrient management, the idea is to precisely target fertilizers and other soil amendments to maximize efficiency, thus reducing the N_2O that occurs as a result of incomplete uptake of the nitrogen-based fertilizers. Another method, conservation tillage, reduces or even eliminates the turning of the soil. This technique prevents the escape of the tremendous volumes of carbon dioxide that are resident in the earth. By improving irrigation practices through any number of methods, farmers can increase the productivity of their crops, another way to avoid the waste of soil carbon. Various techniques also exist to reduce the not-inconsiderable amount of methane that rice paddies generate — six hundred million tons a year of CO_2eq. Agroforestry — the raising of livestock and/or crops on land that also grows trees—is a way to enhance the biosequestration of carbon in the soil. Allowing cropland to revert to its original state, grasslands in some cases and wetlands too, allows for the rapid recovery of soil carbon.[96]

In the United States, the Conservation Reserve Program administered by the Department of Agriculture has, since it began in 1985, helped reduce soil erosion by 622 million tons, restored biodiversity by restoring natural habitat for wildlife, and brought back over two million acres of wetlands, while also increasing the ability of the soil to retain millions of tons of carbon dioxide.[97]

There are some three thousand edible plant species. Unfortunately, 80 percent of the world's farmlands are planted in just ten annual grains, legumes, and oilseeds. Wheat, rice, and maize—what we know as corn in the United States—cover half of the world's croplands. Because these are annuals, needing replanting every growing season, soil carbon is constantly being released, and the gargantuan inputs of pesticides and fertilizers required to cultivate the crops degrade the soil. Water inputs are also a constant concern. Perennials don't require so many inputs, and they don't release soil carbon through their cultivation. Because their root systems grow so deeply and are so extensive, they retain inputs of water and nutrients much more easily, and they hold soil. Researchers are well advanced in producing perennial versions of key crops such as rice, sorghum, wheat, rye, and sunflower.[98]

Farming for the Earth—and Its People

Anna Lappé is the author of *Diet for a Hot Planet* and runs, along with Frances Moore Lappé, her mother, the Small Planet Institute. Lappé *mère* wrote the classic *Diet for a Small Planet* in 1971 and is still going strong today. Her book emphasized the efficacy and the equity in an approach to agriculture that was geared primarily to a vegetarian diet. It was written long before we had as clear a view as we do today of the climate implications of our hyper-industrialized agriculture. *Diet for a Hot Planet* gets to the heart of that concern.

In the book, Lappé *fille* visits a 106-acre farm in Wisconsin that is a dream of sustainability. It has all the five ingredients necessary for her recipe for climate-friendly farming: it respects and follows the path of nature; it restores the earth; it is regenerative in that it mitigates climate change; it is resilient in that it holds water and soil, retains nutrients, and is pest-free with virtually no pesticides or fungicides; and the farm is community oriented, being a member of a cooperative with a thousand members. This idyll, producing very real—and profitable—food, promotes sustainability of the earth and the economy.

Lappé asserts that sustainable agriculture, and horticulture for that matter, are not fads. "Just in the past few years there are all these indicators that have shown just how explosive the urban agriculture and backyard gardening movement has been in this country. Organic seed companies, for example, have been completely selling out." She also notes that the appointment of a world-class leader in eco-agriculture like Kathleen Merrigan to be deputy secretary of the U.S. Department of Agriculture in the Obama administration is a testament to the growing power and voice of the environmentalists and progressive farmers and gardeners who are driving this movement in the United States.[99] Merrigan was one of the authors of the Organic Foods Production Act of 1990 and created the "Know Your Farmer, Know Your Food" program when she came to the USDA in 2009.

The Organic Farming Research Foundation (OFRF), founded in 1990, is one of the leaders in the field. It is involved in sponsoring research, teaching farmers and consumers, developing policy, and advocating for more research and education. OFRF maintains the Organic Farmers Action Network, which keeps farmers in the loop, and is developing a Scientific Congress on Organic Agricultural Research.[100] How many organic farmers are there in the United States? As of 2008, nearly fifteen thousand, in all fifty states. Twenty percent of those were in California, which accounted for better than a third of all the sales. Organic farms averaged 61 percent better sales figures than conventional farms.[101] Globally, as of 2009 there were more than 1.2 million farmers using organic practices, most of them in the developing world.[102]

Productivity and profitability are a big issue, as one would imagine. One seminal study looked at the results from 286 sustainable agricultural interventions in fifty-seven countries over the course of a dozen or more years. The average crop yield increase was 79 percent, with all the farms showing gains in water-use efficiency. Seventy-seven percent of the projects for which there were data on pesticide use showed an average decrease in pesticide use by 71 percent, while yields grew by 42 percent. Fewer chemical inputs mean not only less environmental pollution but also significant cost savings for farmers.[103]

In 2002, the World Bank and FAO undertook a full-scale assessment of world agriculture, the International Assessment of Agricultural Knowledge, Science and Technology for Development. The study culminated in a peer-reviewed report published in 2008 that reflected the work of four hundred experts. Prominent among its conclusions was the fact that although food production per capita had been increasing since the early 1960s, undernourishment of hundreds of millions of people in developing nations persisted. In addition, as the director of the assessment noted in a presentation at the launch, "Productivity increase has come at a cost: environmental sustainability — soils, water, biodiversity, climate change."[104]

The report itself is clear: "In order to increase farmers' natural capital and thereby increase long term flows of farm outputs, modifying the management of soil, water and vegetation resources, based on agroecology, conservation agriculture, agroforestry and sustainable rangeland and forest management, as well as wildlife biology and ecology has been supported."[105]

So, can a significant deviation from the industrial agriculture of the present feed the seven billion human souls on the planet now, not to mention the ten billion of us who will be here at the end of the century? The UN special rapporteur on the right to food, Olivier De Schutter, along with twenty-five experts on agroecology, announced at a seminar in June 2010 that "agroecology outperforms large-scale industrial farming for global food security." De Schutter was unequivocal: "Today, most efforts are made towards large-scale investments in land — including many instances of land grabbing — and towards a 'Green Revolution' model to boost food production: improved seeds, chemical fertilisers and machines. But scant attention has been paid to agroecological methods that have been shown to improve food production and farmers' incomes, while at the same time protecting the soil, water, and climate."[106]

Anna Lappé quotes the farmer Mark Shepard, from the extraordinary operation in Wisconsin that she showcases, New Forest Farm, referring to changing the existing agricultural paradigm: "We can turn it around and we can turn it around fast."[107] Lester Brown echoes this sentiment in a segment of a television series, *Journey to Planet Earth*, that focuses on his book *Plan B 4.0: Mobilizing to Save Civilization*. Brown recalls how quickly after Pearl Harbor, and how thoroughly, Franklin Roosevelt caused the industrial power of America to transition from consumer goods to a war economy. In the TV program, former U.S. secretary of the interior Bruce Babbitt recalls the lightning speed with which the civil rights movement took hold in the early 1960s and how it produced transformative change before the decade was out. He also notes how the best ways to confront the climate crisis have become a subject for intense discussion and debate. "Those debates are happening everywhere clear down to the grassroots in a wonderfully positive way that will yield national change."[108]

Low Tech

We looked at an array of high-tech and low-tech solutions on energy in Chapters 2 and 3. Among these was using biomass for biochar production to achieve a number of important aims. Another low-tech solution that is increasingly being deployed and for which there is growing support is the simple transition to cleaner cookstoves. Because of the burden of soot that is given off by open hearths and badly designed cookstoves around the developing world, scores of millions of people, most often women and children, have much higher incidences of lung disease. Gathering biomass, wood and dung, for cooking leads to significant degradation of forest cover or the waste of a valuable soil nutrient, and has human security implications. Women who gather biomass far from their villages are too often subject to physical violence.

The Global Alliance for Clean Cookstoves is a public-private partnership announced in September 2010 and led by the United Nations Foundation, with the substantial participation of Shell Oil, the governments of the United States, Germany, Norway, Denmark, and Malta, as well as other entities like the World Bank and Morgan Stanley. Its aim is to provide 100 million homes with efficient, clean stoves by 2020. U.S. secretary of state Hillary Clinton, a driving force in this initiative, said in announcing the new alliance, "Clean stoves could be as transformative as bed nets or vaccines."[109]

Another important consideration is that the black carbon that is given off by the fires and stoves contributes in no small way to the problem of climate change. The transportation and deposition of black carbon has a large warming effect. In Central Asia, most of the black carbon that is a major contributing factor to the melting of the Himalayan glaciers comes from cooking fires.[110]

One of the technologies that has been very successful in addressing the multiplicity of problems associated with inefficient and dirty cookstoves is solar cookers. Using the considerable heat of the sun available in many of the developing countries that are trying to meet this challenge, solar cookers have been very successful. Solar Cookers International, one of the leaders in the efforts to bring this approach to communities throughout the world, estimates that hundreds of thousands of solar cookers have been built and are being used. Solar cookers are a good, proven way to cook more nutritious food; avoid pollution; obviate the need to gather biomass, thus saving time, money, and the local environment, as well as being safer for the women who do the gathering; heat and sterilize water, thus significantly lowering the risk of waterborne diseases; reduce the expense to governments of fossil fuel subsidies; and provide business opportunities for families.[111]

Distributed Agriculture

We have seen how the themes of community empowerment and sustainable development have been repeated and amplified in a number of ways in looking at forests and farms. These principles are no less applicable in our cities than they are in rural areas. They are at work in the developed world as well as in the developing countries. Powerful currents have been moving to bring the key principles of sustainability to our urban neighborhoods, creating businesses and livelihoods, improving health and social welfare, educating kids, and fostering community values of cooperation, understanding, and celebration.

In Chapter 3, we saw the promise of the distributed generation of power. DG, or DE — for decentralized energy — is making energy available to local consumers without all the costs, waste, and pollution of gigantic central generating facilities and their attendant transmission and distribution grids. (Remember that electrical power plants using coal, oil, or nuclear typically lose two-thirds of the energy input to the facility to waste heat. This is the same proportion of energy loss in an internal combustion engine.)

Nevin Cohen, an expert on urban planning and food systems, calls urban farming and gardening "distributed agriculture." Dr. Cohen asks the question: "If we have only patches of very small space in cities, can we think of a way to design infrastructure, supporting businesses and distribution systems to make it farmable, to aggregate enough small spaces into something that is profitable and productive?" Stipulating that large urban environments cannot be entirely "off the grid" of the larger production and distribution systems that exist to feed the world, still there are a range of good reasons to create distributed agriculture systems. "The benefits of urban agriculture are not just in the food production but in all the other social and economic benefits as well: managed public open space, youth development, senior social services, and other really interesting programs that get developed around community gardens and urban farms. That includes selling food."[112]

Will Allen is an urban farmer who has carried his model beyond Milwaukee, where his organization, Growing Power, is based. He still runs a thriving program there, though, based on the concept of a "Community Food System." Much of the work of Growing Power is oriented to educating and involving youth. The goal, as he says, is a simple one: "to grow food, to grow minds, and to grow community." Allen received a MacArthur "genius grant" in 2008, consults with urban agriculturalists around the world, including working with the Clinton Global Initiative to develop programs in Africa, and Growing Power itself now has training centers in partnership with local groups in six American cities, beyond its own farms and training in Milwaukee and Chicago.[113]

Allen's daughter, Erika, put it this way: "Overall, it's about helping people use their resources to build soil and grow food."[114]

These are also the sentiments and animating vision of La Via Campesina, an international movement with 150 local and national organizations in seventy countries representing about two hundred million farmers. It vigorously defends small-scale eco-agriculture. Its website declares that "it strongly opposes corporate driven agriculture and transnational companies that are destroying people and nature."[115] Not surprisingly, Anna Lappé identifies it as one of the most important and effective voices in sustainable agriculture in the world today.

These voices are proliferating and gaining strength. In the United States, the growth of urban, backyard, and smallholder farms evokes an earlier time. Lappé recalls the importance of the World War II "Victory Gardens" from which Americans derived 44 percent of their non-protein food needs.[116] The U.S. Department of Agriculture's "Know Your Farmer, Know Your Food" is a modern-day commitment to strengthening local and regional food systems. Some statistics are useful: farmers markets have tripled since the mid-1990s to over six thousand today; there are now over four thousand community-supported agriculture operations in which consumers make commitments to buy from local farmers; there are nearly twenty-two hundred farm-to-school programs in forty-eight states.[117]

USDA is also working to support what Nevin Cohen identified as a crucial need: good ways to transport food from small and medium-size regional farms, and urban farms, and a place to market it. In other words, infrastructure. In the parlance of farming, these are known as "food hubs." USDA deputy secretary Kathleen Merrigan recognizes the need. "Food hubs are innovative business models emerging across the country specifically to provide infrastructure support to farmers. While food hubs are a nascent industry, and many operational food hubs are less than 5 years old, they are based on a time-proven business model of strategic partnerships with farmers, distributors, aggregators, buyers and others all along the supply chain."[118]

The RUAF Foundation maintains an international network of resource centers on urban agriculture. (RUAF stands for Resource Centre on Urban Agriculture and Food Security.) RUAF operates primarily in twenty cities, working with local partners including governments, producers, universities, and NGOs. Its reach is from Bogotá to Belo Horizonte, from Accra to Capetown, and from Amman to Hyderabad to Beijing. With more and more of the world's population headed for our cities, food security is of primary importance. RUAF is trying to build this into how our cities plan and grow and also to incorporate some of the other values we have mentioned. Here is how they see it: "Next to food security, urban agriculture contributes to local economic development, poverty alleviation and social inclusion of the urban poor and women in particular, as well as to the greening of the city and the productive reuse of urban wastes."[119]

Of course, eating healthier food is a message that is incorporated in most of these programs and movements, rural and urban, in the developed world and in emerging economies. Education about health and nutrition is essential. Both Nevin Cohen and Anna Lappé have high praise for Michelle Obama's "Let's Move" initiative to get American kids eating healthier diets and exercising more. The importance of Michelle Obama's attention to the issue is "huge," according to Lappé.[120] "Let's Move" fully embraces the need for children to be eating far more fruits, vegetables, whole grains, and low-fat dairy products. They are also, along with their parents and their communities, being taught about nutrition, farming and gardening, and the value of an ecological ethic.[121]

Educating consumers about what foods have been raised in a sustainable fashion is one of the missions of organizations like the Rainforest Alliance and the partnership in which they are a founding member, the Sustainable Agriculture Network (SAN). The Rainforest Alliance certifies over one hundred crops according to the standards of SAN. As of 2011, 85,529 small and large farms and cooperatives in nineteen countries in the Americas, Africa, and Asia had been certified. The crops include key commodities such as bananas, chocolate, coffee, flowers, juice, mangoes, and tea. Retail outlets, restaurant and hotel chains, airlines, and other venues that sell these certified products are in the major developed economies of the United States, Japan, Germany, the UK, Italy, Holland, Canada, and some others.[122]

SAN, perhaps needless to say, reflects the same values we have seen evidenced throughout this part of the story of how the world is addressing the critical issues of climate change and sustainable development. Its goal is nothing less than "to transform the environmental and social conditions of agriculture through the implementation of sustainable farming practices."[123]

The success stories are sometimes amazing, very often heartening if not downright inspiring, and piling up every day. Even dedicated, well-traveled experts like Jeff Hayward of the Rainforest Alliance take notice of the pace of positive change. In one key area, certification, Hayward noted, "the growth has been astonishing."[124] There are, in short, many very hopeful portents for the future.

Tensie Whelan, speaking for her organization, might just as well be speaking for the movement. "The Rainforest Alliance's work to transform land use and business practices has been in the forefront of the effort to tip the balance decisively towards a sustainable future. Today, the scales are moving our way."[125]

a resilient future

ADAPTATION, EDUCATION, LAW, AND LIFESTYLE

Impacts

The Intergovernmental Panel on Climate Change has produced four Assessment Reports. The First Assessment Report came out in 1990 and played a tremendous role in galvanizing world opinion for addressing the climate crisis. The fifth will be rolled out from September of 2013 through October 2014. Each of the four Assessment Reports to date have had, as you would imagine, a large amount of material on the impacts of climate change. The reports have become increasingly more thorough — and consistently more dire in what is being seen by thousands of scientists all over the planet.

Figure 8.1 shows a chart from the *Stern Review on the Economics of Climate Change*.[1] It is illustrative of the global breadth of the impacts from climate change. As you can see, the impacts intensify as the temperature rises.

A similar scope of impacts is reflected in the study published in 2008 from the National Research Council, one of the National Academies of the United States, on the "Ecological Impacts of Climate Change." The report notes that "climate change is transforming ecosystems on an extraordinary scale, at an extraordinary pace. As each species responds to its changing environment, its interactions with the physical world and the organisms around it change too. This triggers a cascade of impacts throughout the entire ecosystem."[2]

The NRC study, like the compilation from the IPCC, looks at a vast range of impacts, both observable over the course of the recent past and for the future. The IPCC looks at the impacts on a range of sectors, including freshwater resources; ecosystems; food, fiber, and forest products; coastal systems and low-lying areas; industry, settlement, and society; and health. The IPCC Working Group II (WG II) report also zeroes in on each of the continents, as well as the polar regions and small islands.

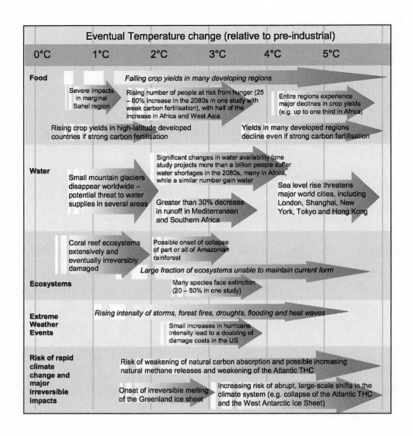

Eventual Temperature change (relative to pre-industrial)

| 0°C | 1°C | 2°C | 3°C | 4°C | 5°C |

Food

Falling crop yields in many developing regions

Severe impacts in marginal Sahel region

Rising number of people at risk from hunger (25 – 60% increase in the 2080s in one study with weak carbon fertilisation), with half of the increase in Africa and West Asia.

Entire regions experience major declines in crop yields (e.g. up to one third in Africa)

Rising crop yields in high-latitude developed countries if strong carbon fertilisation

Yields in many developed regions decline even if strong carbon fertilisation

Water

Small mountain glaciers disappear worldwide – potential threat to water supplies in several areas

Significant changes in water availability (one study projects more than a billion people suffer water shortages in the 2080s, many in Africa, while a similar number gain water

Greater than 30% decrease in runoff in Mediterranean and Southern Africa

Sea level rise threatens major world cities, including London, Shanghai, New York, Tokyo and Hong Kong

Ecosystems

Coral reef ecosystems extensively and eventually irreversibly damaged

Possible onset of collapse of part or all of Amazonian rainforest

Large fraction of ecosystems unable to maintain current form

Many species face extinction (20 – 50% in one study)

Extreme Weather Events

Rising intensity of storms, forest fires, droughts, flooding and heat waves

Small increases in hurricane intensity lead to a doubling of damage costs in the US

Risk of rapid climate change and major irreversible impacts

Risk of weakening of natural carbon absorption and possible increasing natural methane releases and weakening of the Atlantic THC

Onset of irreversible melting of the Greenland ice sheet

Increasing risk of abrupt, large-scale shifts in the climate system (e.g. collapse of the Atlantic THC and the West Antarctic Ice Sheet)

8.1 Impacts from rising temperatures. *Stern Review*

What are these impacts? More and bigger glacial lakes; the thawing of permafrost; earlier snowmelt in mountain regions; warming of lakes and rivers; earlier timing of biological events in spring such as bird migration; a shift toward the poles for terrestrial animal and plant species; shifts also in the range and the abundance of vital ocean species such as algae, plankton, and fish; effects on crops and forests; health impacts from heat, disease, and allergies.[3] The list goes on beyond these.

Glaciers

In Chapter 6, we touched on how black carbon deposition affects glaciers and ice in the Arctic. We did not delve into how thoroughly diminished the world's cryosphere* has become over the past century, with an acceleration over the past several decades.

The World Glacier Monitoring Service and the UN Environment Program issued a comprehensive report on the state of the world's glaciers in 2008. In it they documented the overall trend of the retreat of glaciers worldwide — with a doubling of the average annual melting rate over the last decade! The report predicts the disappearance of glaciers from many mountain ranges altogether during the twenty-first century if measures are not taken to reverse global warming and other negative influences on the cryosphere. In sum, the authors say that "changes in glaciers and ice caps provide some of the clearest evidence of climate change. . . . These changes have impacts on global sea level fluctuations, the regional to local natural hazard situation, as well as on societies dependent on glacier meltwater."[4]

Let's look at the last of these three first: the importance of glacial sources of water for human populations, their agriculture, and their energy. Glaciers are built from fallen snow, in some places accumulated over millennia. They are distinguished by the fact that they move, like very slow rivers, toward the ocean. Glaciers cover 10 percent of the world's land surface — down from the last ice age, when ice covered 32 percent of the land. Locked up mostly in the polar regions and at very high altitudes, they account for 75 percent of all the world's freshwater.[5]

After the Arctic and Antarctic regions, the Himalayas, sometimes known as the "Third Pole," have the largest concentration of glaciers. What distinguishes the Himalayan glaciers from those of the Arctic and Antarctic is that they are the water source for hundreds of millions of people in Asia. There are, by one count, over forty-six thousand glaciers in the Tibetan Plateau.[6] They feed Asia's greatest rivers: the Indus, the Ganges, the Brahmaputra, the Irrawaddy, the Mekong, the Yangtze, and the Yellow.

As with most of the world's ice and glacial masses, the Himalayan glaciers are retreating at an alarming rate. The World Glacier Monitoring Service estimated in 2005 that 90 percent of the glaciers they surveyed worldwide

---

* According to the IPCC, the cryosphere is "the component of the climate system consisting of all snow and ice (including permafrost) on and beneath the surface of the Earth and ocean." Permafrost is defined as "perennially frozen ground that occurs where the temperature remains below 0°C for several years."

were retreating. Lonnie Thompson, one of the world's most distinguished glaciologists, has said that 95 percent of those surveyed in the Himalayas are shrinking. In fact, Thompson has said that the pace of change is speeding up. "Those changes—the acceleration of the retreat of the glaciers and the fact that it is a global response—is the concerning part of all this. It is not any single glacier."[7]

The IPCC used numbers from the *Stern Review on the Economics of Climate Change* saying that three-quarters of a billion people in Asia would be affected, leading eventually to massive water shortages for both drinking and for irrigation. The IPCC puts it this way: "Expansion of areas under severe water stress will be one of the most pressing environmental problems in South and South-East Asia in the foreseeable future as the number of people living under severe water stress is likely to increase substantially in absolute terms."[8]

Asia is increasingly moving to build hydropower facilities, and these would also be impacted by substantially less flow because of the thawing upstream on the Tibetan Plateau. In India, half of the hydropower is dependent on the Himalayan glaciers. Pakistan also uses hydropower, generated from the Indus, which has its headwaters in the Himalayas. China is enormously reliant on hydropower. Similarly, several South American countries use massive amounts of glacier-fed water for their electricity. Hydro accounts for 81 percent of Peru's electricity, 73 percent of Colombia's, 72 percent of Ecuador's, and 50 percent of Bolivia's, much of the water coming from Andean glaciers. In Europe, 50 percent of Switzerland's electricity is fed by glacial waters driving hydroelectric facilities. In Austria, 70 percent comes from hydro, much of it glacier fed. In North America, British Columbia gets 90 percent of its electricity from hydro.[9] The diminishment of glacial mass is a looming threat for power producers in several parts of the world.

Another phenomenon in this context is the creation of serious hazards to populations below the melting glaciers. As a glacier melts, it can significantly build up glacial lakes that accumulate, in many cases, behind unstable rock debris. The natural dams are inherently unstable, and so too much pressure from the water, or earthquakes, or even a wave generated by falling rock, ice, and/or snow can cause them to breach, sending potentially catastrophic discharges of water below. These are called glacial lake outburst floods (GLOFs). They can knock out infrastructure, like hydroelectric dams, and whole communities. In the Himalayan areas of India, Nepal, Tibet, and China, these events have been on the increase.[10]

The United Nations Environment Program represents the problem this way: "Parallel to the worldwide glacier retreat, numerous glacier lakes have been forming at a rapid rate — especially on the surface of debris-covered glaciers.... Therefore, the number of hazardous glaciers, where outburst floods endangers human life and resources, is rising."[11]

Sea-Level Rise and Storms

One of the more salient aspects of the melting of glaciers worldwide is the effect it is having on sea-level rise. Sea-level rise is driven, more or less equally, by two factors: the melting of the earth's ice and the thermal expansion of the oceans by temperature increases. As the thawing of the cryosphere accelerates and temperatures continue to rise, sea-level rise will accelerate as well. There is a broad spectrum of possibility over the course of the rest of the century, but under no scenario is there a lowering of sea level.[12]

One seminal paper put it this way: "Thermal expansion of the warming ocean provides a conservative lower limit to irreversible global average sea level rise of at least 0.4–1.0 m. . . . Additional contributions from glaciers and ice sheet contributions to future sea level rise are uncertain but may equal or exceed several meters over the next millennium or longer."[13]

With higher sea-surface temperatures come more intense storms. Higher sea levels and more frequent and higher storm surges exacerbate the vulnerability of well over a hundred million people worldwide who are subjected to the stresses of tropical storms every year, many of them in the mega-delta regions of South Asia, Southeast Asia, and East Asia. In the two decades from 1980 to 2000, over 250,000 deaths were associated with tropical cyclones, a staggering 60 percent of these in Bangladesh. The deadliest storm in history took place in 1970 and left three hundred thousand to half a million dead in East Pakistan (now Bangladesh) and India's West Bengal region.[14]

Aside from mortality, there are health impacts associated with intense tropical storms and sea-level rise. Catastrophic flooding induces injuries, infectious disease, and trauma-related mental health problems, along with the loss of medical-care personnel and infrastructure. Saltwater intrusion into freshwater impacts drinking-water supplies and agriculture. Cropland is lost to flooding, and so food supplies may be seriously diminished. There are, of course, massive economic costs as well because of the destruction of or damage to habitat, roads, power stations, and other infrastructure.[15]

In the future, there will be significant "hot spots" around the world that will be particularly vulnerable to sea-level rise and storms. These will be places like the Mississippi Delta, where natural and manmade changes have been wrought over decades. These changes have lowered nature's barriers to the sea's potential for causing havoc — as was proved with Hurricane Katrina. Other areas that will be greatly impacted include Australia's Great Barrier Reef, weakened by other climate change impacts such as ocean acidification and higher water temperatures than usual; the Asian mega-deltas, where there are huge populations in the coastal floodplains; those small islands and other coastal areas where freshwater may be overwhelmed by saltwater; the many beautiful and attractive garden spots where tourism thrives; those places where migration to inland areas is not necessarily an easy path; and those parts of the world where adaptive mechanisms are not easily deployed, either for lack of financial and technical resources or because the obstacles are too large to overcome.[16] Impacts from the rising and stormier oceans are going to be a significant challenge, in nearly every case, to overcome.

Sea-level rise, more intense tropical storms, avalanches and glacial lake outburst floods from shrinking ice and snow, and severe stresses to drinking and irrigation water are all phenomena that are upon us now. These and a whole range of other climate change impacts, from drought in agricultural lands in more arid regions to increasingly devastating inland storms such as those that struck the U.S. Midwest and South in the spring of 2011, and then upstate New York and Vermont later that same year, are all trending toward what many are calling the "new normal."

Even with all the Herculean efforts that are under way, from renewable energy to green building, from programs to prevent deforestation and promote sustainable agriculture, to the heartening developments in business and finance, and with the increasing, steadily accumulating political will in nations, cities, states, and regional groupings in Europe, the United States, and elsewhere, we are still going to be under the specter of worsening climate change impacts for some time to come. The same important paper that was referenced above on sea-level rise also calculates that "the climate change that takes place due to increases in carbon dioxide concentration is largely irreversible for 1,000 years after emissions stop."[17] What we can influence, of course, is the severity of the changes. We can mitigate the amount of the greenhouse gases we put out and the other forces, like deforestation, that are exacerbating climate change. In the meantime, it is incumbent upon us to learn how to adapt.

Heidi Cullen is a prominent research scientist, author, educator, and the head of Climate Central, a clearinghouse for accessible information on climate change. In her book *The Weather of the Future*, she catalogs a number of developing and developed world scenarios for adaptation — or what might happen in the absence of adequate adaptation strategies. She quotes Mark Twain: "It's better to build dams than to wait for the flood to come to its senses."[18]

The Insurance Industry's Response

One industry that knows what is coming is insurance. We looked at insurance in Chapter 5, noting how fully convinced all the leaders in the industry are of climate change's impacts, now and for the future, and how committed they are to managing those risks through pushing for policy to meet the crisis head-on and by promoting measures to effectively adapt to the inevitable stresses on human populations and infrastructure that are at hand. The Insurance Information Institute cites one study on hurricanes that indicates that as wind speeds increase over the next couple of decades, property insurance losses will increase as well — by 30 to 40 percent. Seven of the ten most costly hurricanes in U.S. history occurred from August 2004 to October 2005, including Katrina, which caused losses of $41 billion. If the predictions of more-intense storms bear out, with the attendant increases in property loss, then Katrina will be dwarfed in the future, especially if hurricanes zero in on major cities in the United States like Miami or New York. The Big Apple had a near miss with disaster in late August 2011 with Hurricane Irene.

Because of the risks, insurers have raised the ante for companies and individuals. The threshold for capital adequacy for companies is much higher, and the risk from losses from two disasters happening in quick succession must be considered.[19]

Beyond this, insurers are simply pulling out of high-risk areas. Allstate, the third-largest player in the United States, has been reducing its exposure in hurricane zones and in areas subject to cyclones and other inland storms. Allstate's CEO, referring to more-intense storms, has said: "We're running our business as if this change . . . is permanent. We don't know whether the genesis of that increase over the last three years is just a cyclical movement in weather, whether we're to 'normal' or we were at 'normal' before, or it's global warming or anything else, but we're acting as if this will be permanent and we need to recover those costs."[20]

As terrible as loss of life and property damage may be for communities in the developed world, it is exponentially worse for those in the developing world. These populations often have minimal shelter, limited food and water resources, poor or, in many cases, nonexistent infrastructure, and, what is a particularly pernicious factor in a catastrophe, inept, corrupt, and indifferent governance. Witness the disaster in Myanmar (Burma) in 2008 that ensued from the landfall of Cyclone Nargis and the resistance by the military government to offers of aid by the international community. Certainly the lack of preparedness and the meager response by the junta as well as its initial blocking of outside aid contributed to the massive loss of life that numbered near 140,000.

In any event, the insurance industry is among those who have taken a lead role in helping developing world communities to adapt to the many difficult climate change impacts to which they are subject. In 2005, Munich Re launched the Munich Climate Insurance Initiative (MCII). In 2010, MCII joined with three other groups, with membership among them of over one hundred major international insurance firms. They issued a statement with concrete proposals for governments to adopt in order to help developing countries adapt. The insurance companies want to "catalyse adaptation efforts through risk management, loss prevention and risk transfer." They offer various successful models such as the Caribbean Catastrophe Risk Insurance Facility and Horn of Africa Risk Transfer for Adaptation as indications of how effective these sorts of approaches can be. The companies have made a commitment to provide their expertise and to deliver information and advice on options, developing new insurance products, and raising awareness among stakeholders.[21]

Ceres, the investment responsibility group, looked at the industry a few years ago and even then counted over six hundred climate initiatives ongoing at nearly 250 companies worldwide.[22]

The Need to Adapt — and the Money with Which to Do It

Adaptation is universally recognized as a fundamental strategy for all the communities of the world that are going to be affected by climate change — and that is all of us. The parties to the UN Framework Convention on Climate Change have developed a range of approaches. They have long since delineated the risks and recognize what needs to be done: "Increased investment in adaptive capacity, such as strengthening the ability of countries to reduce disaster risk, will safeguard economic progress already made and increase the climate resilience of economies on the way to achieving overall development goals."[23]

In 2005, the UNFCCC launched the Nairobi Work Program to help parties, particularly those in the developing world, understand impacts and the options for adaptation. Its five-year mission brought together expertise from 207 partners in civil society, the private sector, UN agencies, research institutes and academic institutions, and the like. They performed work on everything from capacity building to disaster risk reduction, and from enhancing communications and monitoring capabilities to performing "climate-resilient development planning." These entities brought their resources into focus on the full range of climate hazards, including drought, floods, heat waves, wildfires, and tropical storms.[24]

In 2001, the parties recognized the need to assist the Least Developed Countries (LDCs) in adaptation planning and implementation. In order to effect this, a program of support for National Adaptation Programs of Action (NAPAs) was created. Forty-five NAPAs have been written, and nearly five hundred projects have been identified within them.[25] Implementation of the projects is ongoing, with some funding, $180 million, having been provided through the Global Environment Facility's Least Developed Countries Fund. It aims to raise another $500 million in the near future to continue the program and expand it.[26]

Another key funding mechanism is the Adaptation Fund, established by the UNFCCC with the World Bank as the trustee for the money. The principal source of income for the fund is from a share of the proceeds from the sale of Certified Emission Reductions under the Clean Development Mechanism. By the end of 2012, the fund should have over $350 million in hand.[27]

These are good numbers for a relatively young program, and it will expand. Meanwhile, of the climate funding received as of early 2012, nearly 25 percent has been earmarked for adaptation: a little over $400 million.[28]

But the real money is going to come from what has been termed "fast-start finance." Prior to the dramatic meetings in Copenhagen in December of 2009, there again, and subsequently at Cancún in December of 2010, the developed world nations made a commitment to provide funding of $100 billion a year by 2020 for both mitigation and adaptation in the developing world. For the period 2010 through 2012, $30 billion was pledged.

After Copenhagen, the UN secretary-general convened a High-Level Advisory Group on Climate Change Financing (AGF), which met over the course of the year and made its report to the secretary-general prior to Cancún. The AGF had some high-profile actors such as George Soros, Nicholas Stern, Larry Summers, Christine Lagarde (now the head of the International Monetary Fund), and Chris Huhne (then the UK's climate and energy minister), among a number of other international financial worthies. In their final report, they wrote, referring to the $100 billion goal, that "the Advisory Group concluded that it is challenging but feasible to meet that goal. Funding will need to come from a wide variety of sources, public and private, bilateral and multilateral, including alternative sources of finance, the scaling up of existing sources and increased private flows. Grants and highly concessional loans are crucial for adaptation in the most vulnerable developing countries, such as the least developed countries, small island developing States and Africa."[29]

As of the spring of 2011, developed countries and the EU, apart from its member countries' contributions, had pledged slightly over $28 billion for fast-start finance. Japan has led the way with its commitment of $15 billion. The funds, when they are finally allocated for specific projects, are supposed to reflect a balance between adaptation and mitigation.[30]

At Cancún, a Green Climate Fund (GCF) was designated as the administrator for the money. A transitional committee was appointed to make recommendations for the design of the GCF and to report at the Seventeenth Conference of the Parties in Durban, South Africa, at the end of 2011. The GCF was officially "launched" in Durban, but because of the considerable amount of money involved, the need for the utmost probity in the design and the operations of the GCF, and the complexity of the politics involved between the many stakeholders, it will take a number of months, stretching into 2013, before the GCF is fully operational.[31]

How to Adapt: Holding Back the Sea

The IPCC, like the UNFCCC and scores of development and relief agencies, NGOs, and think tanks, long ago recognized the pressing need to build the capacity to adapt to changing conditions. The IPCC has cataloged some of the ongoing adaptation methods and strategies and gauged costs and benefits. It has looked, generally, at various adaptation options for different circumstances and for different societies, recognized some of the constraints, not the least of which are financial, and also recognized that "adaptive capacity is uneven across and within societies."[32]

We are, nevertheless, building that capacity, project by project, and gaining expertise in the necessary means of communication, engineering, and finance that will provide communities with more security as climate impacts intensify.

It is, of course, easier to adapt if you have advanced infrastructure and health care systems, and access to financial resources and engineering expertise. "The best starting point for adaptation is to be rich. It is not foolproof: not even the rich can buy off all hazards, and rich countries and individuals will make poor decisions."[33]

In England, the Thames Barrier, originally built by the Greater London Council and now run by the Environment Agency, went into operation in 1982. This phalanx of ten steel gates protects London from storm surges and was conceived after massive flooding inundated the capital in 1953.[34] It is one example of an advanced economy recognizing the need to protect against floods and having the foresight to actually go ahead to protect against the catastrophe of a principal urban area under water. In New Orleans, in 2005 with Hurricane Katrina, it became clear that the various government agencies involved in flood protection hadn't managed their program adequately. In

Bangkok in 2011, flooding swamped the metropolitan region and its twelve million people, with hundreds of lives lost and billions in damages.

In Holland, the Dutch have been fighting for centuries to keep the sea at bay. They know now that their Promethean efforts to date are not going to be enough in the face of continued sea-level rise and the attendant storm surges that will swamp their small country unless they do more. To that end, they have created a Delta Commission to protect the country from flooding and to safeguard drinking water supplies. It is a comprehensive approach, dealing with all the regions of the country, has dedicated funding — earmarked at €1 billion a year by 2020 — and it has the full support of the government and the Dutch people.[35]

It is about fifty-four hundred miles from Holland to California. Despite the distance, these two places have a similar problem: an incredibly complex water control and supply system that is increasingly susceptible to catastrophe because of sea-level rise. In California, you have to add another lurking danger: earthquake. The Sacramento–San Joaquin Delta is the hub for water for two of every three of the Golden State's thirty-seven million people and for four million acres of farmland. Because of higher flows into the rivers feeding the delta and the threat of a storm surge, as well as the specter of an earthquake, this highly engineered system is liable to failure, sooner rather than later, according to many experts. One estimate has it that "the odds are roughly two in three that during the next fifty years either a large flood or a seismic event will affect the Delta."[36]

Water for People, Farms, and Ecosystems

There is another bear trap attributable to climate change waiting to spring: as warming increases in the North American West, the snowpack in the Sierra Nevada mountains is diminishing. So, if a flood or an earthquake does not cripple California's water supply in the medium term, a shrinking freshwater source will in the long term. California water experts predict a 25 to 40 percent reduction in Sierra snow by 2040.[37]

Proposals to deal with the threat to this vital system center on a canal that would be built to take pressure off the delta itself, giving relief to the ecosystems that are increasingly under stress. In any event, water conservation and efficiency efforts are being pushed to the fore, as they must be. This is, of course, a key to adaptation anywhere in the world where water supplies are being stressed.

The Intergovernmental Panel on Climate Change issued a special technical report in 2008 on "Climate Change and Water" in which it looked much more closely at the many ins and outs of freshwater resources than the Fourth Assessment Report had in 2007. The 2008 report emphasized the importance of adaptation and the particular virtues of conservation and efficiency, otherwise known as demand-side management — just as with energy use:

> Adaptation options designed to ensure water supply during average and drought conditions require integrated demand-side as well as supply-side strategies. The former improve water-use efficiency, e.g., by recycling water. An expanded use of economic incentives, including metering and pricing, to encourage water conservation and development of water markets and implementation of virtual water trade, holds considerable promise for water savings and the reallocation of water to highly valued uses. Supply-side strategies generally involve increases in storage capacity, abstraction from water courses, and water transfers. Integrated water resources management provides an important framework to achieve adaptation measures across socio-economic, environmental and administrative systems. To be effective, integrated approaches must occur at the appropriate scales.[38]

Thankfully, this is how most of the world's water managers now see the problem. They see their water supplies diminishing, year in and year out, and they know that planning, educating policy makers and the public, finding financial resources, and building for adaptation is the only reasonable path.

Cities Taking the Lead

Policy makers in world cities like New York also are very serious about the need for adaptation. Much of New York City's key infrastructure is at or near sea level. That includes the city's fourteen sewage treatment plants, much of its subway system, its power and communications systems, and its two major airports. The New York City Panel on Climate Change issued a report that flagged much of this vulnerable infrastructure. A subsequent report looked at the challenges in adaptation for the Big Apple. It recommended, among other things, "Flexible Adaptation Pathways" that would be able to incorporate ongoing research on the best practices to be mounted to safeguard against substantial damage to the trillions of dollars worth of infrastructure at risk.[39] One vision of what New York City will need in the not-too-distant future is a network of four storm surge barriers, thirty feet high, not unlike the Thames Barrier.[40]

Chicago is also planning for hotter and wetter days ahead. In 1995, Chicago's deadliest heat wave killed hundreds of people, mostly elderly. Thousands more were hospitalized. (Over forty thousand died in a brutal heat wave in 2003 in Western Europe, and as many as fifteen thousand died in Russia in the apocalyptic drought, heat, and fires of the summer of 2010.)

Chicago is reconciled to its future and is adapting its infrastructure now. Much of the activity is geared toward reducing the "heat island" effect of the asphalt, concrete, steel, and glass surfaces of the city. Alleys, streets, and parking lots account for 40 percent of Chicago's ground cover. To offset this effect, tree plantings are proliferating on widened sidewalks. Trees provide tremendous benefits in reduced energy use by shading buildings. Chicago has also become a world leader in green roofs. These roofs provide tremendous insulation from the heat and the cold and also serve as buffers for rainwater, slowing down its progress to the storm sewers that have too often become flooded in years past.[41]

One key NGO, ICLEI/Local Governments for Sustainability, has been working in the forefront of efforts to help cities adapt. Their Climate Resilient Communities (CRC) program helps local governments with some of the technical problems that may be new to them or, in some cases, beyond their means to deal with on their own. ICLEI's Adaptation Database and Planning Tool (ADAPT) is an online tool to help planners and policy makers. Its Climate Adaptation Experts Advisory Committee provides key input on programs, and its CRC Steering Committee is continuing to come up with useful programs for cities. Globally, more than twelve hundred cities, towns, counties, and associations are members.[42]

Similarly, the Urban Climate Change Research Network, a group of experts from key institutions like NASA's Goddard Institute for Space Studies, the Institute for Sustainable Cities of the City University of New York, and institutes and universities in the UK, Brazil, Australia, Argentina, and Nigeria, among other places, has produced several reports identifying cities as a key focal point for action and which highlight many of the sorts of action that are already under way.[43] And UN-Habitat has published its *Cities and Climate Change Global Report on Human Settlements 2011.*[44] Beyond its ongoing work, ICLEI has organized two world conferences on cities and adaptation.[45]

The C40 Cities Climate Leadership Group and the World Bank have created a key partnership in which the administrations of many of these forty megacities, the largest in the world, will be able to directly access technical and financial assistance from the World Bank, rather than going through their national governments, the traditional route. Robert Zoellick, World Bank president, said at the launch of the program: "This agreement will help us work with C40 cities to integrate growth planning with climate change adaptation and mitigation, with special attention to the vulnerabilities of the urban poor."[46]

The Public Health

The Centers for Disease Control, the public health agency in the United States, is fully aware of the implications for human health in a changing climate. They have a growing program to address a number of concerns, which include heat exposure from higher temperatures, more intense and more frequent extreme weather events, and an array of medical problems, from asthma and allergies to food-borne and waterborne diseases, among others. Their Climate-Ready States and Cities Initiative aims to anticipate health problems and arm the public health and medical establishments in these jurisdictions with the resources to plan and deal with the many and diverse health problems that have arisen and will become more prevalent.[47]

The U.S. Global Change Research Program (USGCRP), under the aegis of the White House Office of Science and Technology Policy, shares the CDC's concerns regarding the health impacts of climate change. The USGCRP oversees an Interagency Crosscutting Group on Climate Change and Human Health that looks at the many ins and outs of this key area and promotes further research, communicates important findings to all the relevant federal agencies and other stakeholders, and whose "ultimate goal is to build communities that are healthy and resilient to climate change impacts." USGCRP is also deeply involved in adaptation science.[48]

The late Dr. Paul R. Epstein, co-author of *Changing Planet, Changing Health,* and a leading researcher in the intersection of climate change and public health for many years, warned that the impacts were upon us, saying "the consequences of climate change sometimes appear far off. But warming and changing weather patterns are already driving changes in public health." He further warned: "Climate change has multiple implications for human health. Some of the impacts are direct (e.g., from heat waves), and some are indirect (via disease vectors). The most profound implications for human health, however, lie with the impacts of climate change on the ecological systems — our life support systems — that underlie our health and well-being."[49]

Adaptation Planning: A Critical Task

The White House commissioned a Climate Change Adaptation Task Force to produce a comprehensive report with recommendations for how federal agencies should plan for and respond to climate change, and how the federal government can serve communities throughout the country in this regard. The task force's progress report came out in the fall of 2010, and a follow-up came out in the fall of 2011. What the task force is doing is pushing the agenda for climate change adaptation significantly further along than it had been, including developing a consistent, effective strategy for the United States to support international adaptation work.[50]

As noted previously, the impacts of heat, water stress, storms, etc., for less-developed countries are going to be less easily buffered by infrastructure improvements, primarily because of the relative lack of resources. That is why the UN and other organizations are working to help mobilize finance and technical support for the developing world, particularly those Least Developed Countries that are generally the poorest and the least able to adapt.

The U.S. Agency for International Development, for example, has been at work with a number of programs to support adaptation efforts. These have included flood planning for Honduras, securing freshwater in the Marshall

Islands, and food security programs in Mali. USAID has also produced guide-books for planners.[51]

The United Nations Environment Program is very actively involved in helping to create a broad-based adaptation "knowledge base" and practice in the developing world. UNEP has a range of programs across the spectrum: knowledge and policy, science, finance, and "ecosystem-based" adaptation. This is a way of using the "ecosystem services" of natural areas, such as the ability of wetlands and forested areas to protect against floods, to more advantage.[52]

Returning to one of the key impacts we looked at earlier, namely glacier lake outburst floods (GLOFs), UNEP has been identifying ways to minimize the danger. UNEP, working with sister agencies such as the UN Development Program (UNDP) and the World Bank, along with national governments, is involved in an ongoing effort to survey and catalog the threat that GLOFs pose, and to focus efforts on remediation of the threat. One method is to siphon off the buildup of water in these lakes.[53] The International Centre for Integrated Mountain Development, an NGO serving the eight countries of the Hindu Kush–Himalayas, is engaged in mapping and monitoring efforts too.[54]

The Red Cross / Red Crescent Climate Center is yet another key international locus for responding to climate change impacts. These organizations are, of course, world leaders in disaster management, but they are also trying to mitigate the need for full-scale response to catastrophic events by lowering the chances for these to happen in the first place. Early warning systems, communications networks, advocacy, and community risk reduction are all modes that they use to minimize the potential for disaster. As with so many other adaptation programs, building "community resilience" is a key concept.[55]

Expanding and Sharing the Adaptation "Knowledge Base"

The message that we are experiencing unusual, difficult conditions as a consequence of a changing climate, and that we can expect worse before it gets better, is clearly reaching the public and the policy makers. Popular books like *Climatopolis* by noted environmental economist Matthew Kahn, and Heidi Cullen's *Weather of the Future*, are telling the story, eloquently, of what we are in for and how to best adapt. Beyond these books, scientists, economists, and planners have been developing scenarios and strategies to address the impacts. The Intergovernmental Panel on Climate Change has certainly been in the vanguard, and NGOs and multilateral organizations like UNEP and the Red Cross/Red Crescent societies are building adaptation practices. A sampling of books offered by the American Planning Association gives a sense of how seriously the need for adaptation is being taken: books on "delta urbanism" in both New Orleans and the Netherlands; *Planning for a New Energy and Climate Future*; *Post Carbon Cities: Planning for Energy and Climate Uncertainty*; *Sustainable Solutions for Water Resources*; *Planning for Coastal Resilience: Best Practices for Calamitous Times*; and *Planning for the Unexpected*. EcoAdapt, an NGO that is pioneering adaptation options, has produced a reference, *Climate Savvy: Adapting Conservation and Resource Management to a Changing World*.

Technology is rising to the occasion, with new approaches being bruited to help adapt. Flood-prevention engineering is going to grow in importance beyond its already high profile, as are weather and climate forecasting and modeling. Systems to conserve energy and water are also going to be hugely important in the many places where drought is going to be more prevalent. That means technologies for much more efficient irrigation, for desalinization, storage, and recycling are going to be further researched and developed. Genetically modified crops — the debates regarding their safety aside (for our purposes here, anyway) — may prove useful in giving farmers more ways of producing food in a water-constrained environment.[56]

There is a lot of knowledge sharing going on, with online "knowledge bases" and international meetings as well. The International Institute for Environment and Development has organized six conferences through 2012 on community-based adaptation, the last taking place in Hanoi. Australia sponsored a conference in 2010, Climate Adaptation Futures, in which seven hundred delegates took part, learning about the entire spectrum of climate impacts and ways to address them.

Bangladesh and Resilience

Ironically, perhaps, one of the first nations to respond to the catastrophe in Myanmar in 2008 was its neighbor Bangladesh, scene of the worst loss of life in history from a natural disaster. But Bangladesh, arguably the most vulnerable nation on earth to the potential ravages of climate change, has developed a sophisticated set of protocols for dealing with storms and has accumulated expertise in ways to help victims. Bangladesh has become one of the world's leaders in adapting to the vagaries of climate change.

One of the most brilliant adaptations is the development in Bangladesh of the *baira*. These are floating gardens, based on a traditional practice, that enhance the ability of smallholders to grow an assortment of vegetables for both the family and the marketplace while at the same time resisting the destructive properties of the rising waters of the Bay of Bengal. The technique is replicable nearly anywhere in the world and is indeed being studied for use elsewhere. "Cheap, green and effective: what more can one want?" noted one NGO writer.[57]

The *baira* approach illustrates one of the most important themes that we see in how the world's organizations and communities are confronting the climate crisis: sustainability is at the heart of our efforts. We saw this in numerous examples in the last chapter. It is embedded in how adaptation techniques are being developed. As the *Economist* notes: "Action on climate bleeds into more general development measures."[58]

One important concept in this context is "resilience." The UN Development Program is, as you would expect, deeply involved in efforts to adapt to global climate change. It promotes "climate-resilient economic development and sustainable livelihoods." UNDP wants its clients to integrate what they are attempting to achieve in poverty reduction into efforts to minimize risks from climate-related disasters, both the slow-moving kind and those with a rapid onset. Its principal focus is to build the capacity for stakeholders to assess risks, mount effective responses, and to prosper in the bargain. UNDP, as does everyone involved with sustainable development, fully recognizes the many opportunities inherent in building environmentally and economically robust and resilient communities.

Getting Smarter about Sustainability

For the long term, perhaps the most important development that will help enable us to stave off the worst scenarios and, eventually, reverse some of the more troubling trends is we have been educating ourselves. We have been acquiring an unprecedented body of knowledge and gaining practical experience in a vast range of pursuits, from climate science to renewable energy finance, from building political movements and will to building infrastructure to minimize climate impacts, from communication of the many important messages that the public and policy makers need to hear to gathering together in vast networks to promote the work that needs to be done.

As this knowledge has been amassed, more and more educators and students have been clamoring for it and gaining increased access to it. In the 1950s, with the launch of *Sputnik* by the USSR, American politicians and educators felt the need, rightly or wrongly, to boost the levels of education in math and science. Similarly, with the advent of the modern environmental movement, and then the rise of sustainable development, and finally with our concerns about global environmental threats like stratospheric ozone depletion and, above all, global climate change, programs in conservation biology, environmental engineering, ecology, environmental economics, and indeed, sustainability itself, have come into their own.

As we have noted, there has been a slew of traditional media on the various and sundry aspects of climate, energy, and sustainability: books, films, and magazine, newspaper, television, and radio coverage. Beyond that, people are being informed on a daily basis by blogs and websites, and academic papers, journal articles, and studies and reports of all shapes and sizes, accessible via the web and in print. These include all the materials associated with major work like the Assessment Reports, and other studies of the Intergovernmental Panel on Climate Change, as well as profiles of various projects and programs from thousands of NGOs.

Important gatherings, like the annual Conferences of the Parties to the UN Framework Convention on Climate Change, other key events like the Earth Summit in 1992 and the twentieth-anniversary event in 2012, and the hundreds of conventions, workshops, and other forums, large and small, that take place all over the world every year all provide tremendous learning opportunities for the conferees. The subject matter ranges from sustainable development to climate science to finance to green building to clean tech, and scores of other relevant subjects.

The Climate Project is Al Gore's program to train individuals to expertly give the story of climate change, as Gore himself did with his now-famous slide show and the movie *An Inconvenient Truth*. Gore has personally trained over three thousand volunteers, who have given over seventy thousand presentations.[59]

The U.S. Green Building Council, the creator of the LEED (Leadership in Energy and Environmental Design) rating system, has certified more than 160,000 people to be LEED accredited professionals or LEED green associates. These highly trained individuals are the ones who are engaged in the increasingly more prevalent practice of evaluating building projects for designation as USGBC approved in a number of categories and at different levels. A comprehensive curriculum is available, with scores of specialized courses given in person and online, in a short time span or over the course of many weeks. Beyond these training courses, many other courses are available to industry professionals to learn about new techniques and business opportunities. The USGBC also educates building professionals and others at its annual Greenbuild International Conference and Expo, attended in recent years by over twenty-five thousand people.[60]

The World Green Building Council counts almost eighty other GBCs in the world, in various stages of becoming established, most of which have some sort of rating system, a professional education system, and a network of certified professionals.[61]

Other professional organizations like the American Planning Association and the American Institute of Architects, among many others, have ongoing training programs geared to the new realities of a climate-sensitive world.

National Programs

Under the terms of the UN Framework Convention on Climate Change, the developed countries are required to submit "national communications" on their activities. These are a treasure trove of information on the broad sweep of work being done on climate change, including on "Vulnerability Assessment, Climate Change Impacts and Adaptation Measures." Each of the forty Annex I countries — the developed nations of the world — plus the European Union, have submitted these reports through the fifth iteration in 2010. There is a section in each on "Education, Training and Public Awareness."[62]

The Fifth U.S. Climate Action Report (CAR) provides an overview of a wealth of activity taking place in this arena. The authors' thoughtful approach to reporting on "fostering public climate literacy" recognizes that this "will play a vital role in knowledgeable planning, decision making, and governance." The Fifth U.S. CAR catalogs a host of federal agency activities, including the work of the Education Interagency Working Group under the auspices of the U.S. Global Change Research Program. The CAR notes the publication by the National Oceanic and Atmospheric Administration (NOAA) of *Climate Literacy: The Essential Principles of Climate Science*. It is geared to the general public and to educators, as both a guide and a "launching point" for discussion and inquiry. NASA, NOAA, the National Science Foundation, and other federal agencies have all been involved in major outreach programs. These have been geared to the general public, as well as K–12, undergraduate, and graduate students, and professionals within any number of disciplines.[63]

Looking at the chapters on education from the national communications (NCs) of other countries is revealing. Australia's NC5, for instance, notes that "all levels of Australia's education and training system including primary and secondary schools, universities and the vocational education and training sector are embedding the principles, knowledge and skills required to foster long-term sustainability and prepare for climate change."[64] The French report that 81 percent of their citizens polled in 2009 said that climate change is a very serious problem, more than the average of 74 percent at that time among all European citizens. The French have required environmental education and training to be a priority for schools since 2005.[65] The Japanese have a number of education and awareness programs that are geared to increasing environmental education for schools, improving awareness through mass media ads and various public forums, and supporting NGOs in their efforts.[66] This is but a smattering of the many educational programs that are under way in all the Annex I countries.

Reaching the Public: Museums and Continuing Ed

More and more museums and science centers all over the world are communicating the complex stories of climate science, the impacts, and the solutions. The Association of Science-Technology Centers, a group with 541 members in forty countries, created International Action on Global Warming (IGLO) in 2007 to communicate climate science to large audiences all over the world. IGLO has a "toolkit" available to its members and has been cultivating cooperative programs between the centers. Exhibits have popped up from Trento, Italy, to Ottawa, Canada, and from Stockholm, Sweden, to Cape Town, South Africa.[67]

A major exhibition on climate change was produced by the American Museum of Natural History in New York City and ran there from October 2008 through August 2009, then hit the road for Spain, Denmark, Mexico, and Abu Dhabi, as well as Chicago, Cleveland, and Saint Louis.[68] Museums and science centers have had any number of conferences, panel discussions, lecture series, workshops, and the like around climate change. The Smithsonian Institution, for instance, had an online education conference in the fall of 2009, and the Natural History Museum in London has had several "student summits." Again, these examples are merely the tip of the iceberg on these sorts of activities worldwide.

Formal education programs have been proliferating as well. "Continuing education" programs are extremely popular. One large program, New York University's School of Continuing and Professional Studies, lists courses in global climate change, sustainability and corporate social responsibility, and green building and sustainability, among many others. There are, of course, hundreds of schools offering continuing or adult education in the United States alone. "Distance learning" is another vehicle for many of these courses.

Focus on Schools

These sorts of activities and more traditional programs at colleges and universities are appreciably enhanced by the work of professional organizations such as the Association for the Advancement of Sustainability in Higher Education. AASHE has over a thousand member institutions in all fifty states and sponsors a range of professional development programs such as conferences, webinars, and workshops. It also developed the Sustainability Tracking, Assessment & Rating System (STARS), which colleges and universities use to measure their performance in this area.[69]

The U.S. Green Building Council has a Center for Green Schools that has the vision of creating sustainable working and learning environments for everyone. The center has geared its efforts to those schools it has identified as needing the most help: K–12 schools in poorer communities and other "under-resourced" institutions. The center is working on policy with state and local governments and providing practical training and resources to build new schools and renovate old schools that will be greener and, over their life cycles, cheaper to operate.[70]

As with so many successful movements, coalitions are a critical component. The Coalition for Green Schools is an alliance of heavyweights that includes the American Architectural Foundation, the American Federation of Teachers, the National Education Association, the National PTA, the National School Boards Association, the USGBC, and several others, altogether representing more than ten million people. The coalition aims at nothing less than providing every student in America with a green school.[71]

Another highly effective enterprise is the American College & University Presidents' Climate Commitment. This initiative has almost seven hundred signatories, in all fifty states and the District of Columbia, representing one-third of the U.S. higher education population, about six million students. The Presidents' Climate Commitment not only promotes much greener campus facilities and operations, but fosters the continuing growth of sustainability education and research. A big part of the mission is to train teachers for furthering the burgeoning educational universe on sustainability, climate change, clean tech, and the environment, including professionals who will be involved in K–12.[72]

Engineering programs, business schools, and law schools now all offer course work in climate, sustainability, and other "green" studies, many with advanced degrees. You would not be surprised to learn, for instance, that there is a "Teaching Climate Change Law & Policy" blog.

Michael Gerrard, a prolific litigator and writer on environmental law, founded the Center for Climate Change Law at Columbia Law School in 2009. Gerrard wrote *Global Climate Change and U.S. Law*, published by the American Bar Association in 2007. Beyond the course work and the high-level public programming that the center offers, it also maintains a robust website that Gerrard characterizes as the "world's leading reference service on climate change law." The center has attracted many foreign students at Columbia Law, a number of them Australian. It is likely they will go home to work on various aspects of the nascent federal law there that puts a price on carbon.[73]

Information for educators is available from any number of networks. The Climate Literacy and Energy Awareness Network (CLEAN) offers lessons on one "guiding" principle and seven "essential" principles. The guiding principle is that "Humans can take actions to reduce climate change and its impacts." A diverse menu of teaching activities, along with webinars and workshops, is offered. CLEAN was developed by the National Science Foundation, along with NOAA, the Cooperative Institute for Research in Environmental Science at the University of Colorado–Boulder, the Science Education Resource Center at Carleton College, and some others.[74]

The Growing Body of Law

Since the advent of the National Environmental Policy Act in the United States in 1969, and the subsequent landmark laws such as the Clean Air Act, the Clean Water Act, and Superfund, among others, an enormous body of environmental law and lawyers to practice it have grown up. The same pattern applies throughout the world as well and includes international environmental regimes that must often be adjudicated in various venues such as the World Trade Organization and the European Union.

Climate law is a particular specialty within the field, but it has been growing in importance and in the number of cases brought to court. There were 461 climate-related cases filed for litigation in the United States from 1989 through the fall of 2011. As you would expect, the number has been rising in recent years, with 177 filings in 2010.[75] In Chapter 6, we saw how the Sierra Club's Beyond Coal Campaign has been so successful. Much of this success is attributable to the gathering strength of the legal expertise being brought to bear in defense of the climate. Through 2010, there had been forty-six climate cases brought in nine countries outside the United States, plus twenty-three in the European Union courts.[76]

Professor Gerrard of the Center for Climate Change Law understands the big picture. For instance, he attributes the successes in taking coal-fired power plants off the drawing board to "the Sierra Club's really terrific campaign to try to litigate against every plant, number one; and number two, and probably the most important, the declining price of natural gas and its increased availability; three, higher steel prices, largely due to demand from China; number four, the economic recession; and number five, regulatory uncertainty about climate regulation. I think it was all of the above that, combined, led to the new coal-fired power plant industry falling off a cliff."[77]

In that mix, the litigation certainly played a key role. According to Gerrard, "It slowed down a lot of the coal plants, in some cases by years. It made the bankers very nervous. It 'threw a lot of sand in the gears' of regulatory approvals. I do think it played a significant role."[78] Of the 461 climate-related cases in the United States through the autumn of 2011, 22 percent were coal cases.[79]

We saw in Chapter 6 how the U.S. Supreme Court in 2007 not only affirmed the EPA's right under the Clean Air Act to examine whether or not greenhouse gases were an "endangerment," but indeed reminded the EPA of its responsibility under law to make that determination. EPA subsequently found GHGs to be an endangerment, and that is the basis of the regulatory regime that is being built now to curtail their emissions.

In 2011, in another landmark case, the Supreme Court reaffirmed 8–0 that the Clean Air Act places the responsibility for abating greenhouse gases on the EPA. Justice Ginsburg wrote: "It is altogether fitting that Congress designated an expert agency, here, EPA, as best suited to serve as primary regulator of greenhouse gas emissions."[80]

Of course industry has been mounting legal challenges to various EPA rulings. One of the more interesting cases seeking to block greenhouse gas regulation, however, comes from a surprising source: a coalition of environmental justice groups. The Association of Irritated Residents (AIR) et al. sued the California Air Resources Board (ARB) seeking to overturn the board's cap-and-trade regime. The crux of AIR's argument is that cap and trade, because of its flexible approach in allowing polluters to offset emissions, does not require actual reductions. AIR's argument is that economically disadvantaged communities will thus not be protected.[81] The California Supreme Court allowed ARB to continue in its rule-making, and so it issued the final cap-and-trade regulation in October 2011. AIR continues to challenge the law, but in the meantime California is moving ahead with its landmark program.

Professor Gerrard sees the immediate future of the legal landscape in terms of various courses of action.

I think we will see environmental groups pressing claims against some big corporations for lying about climate science. ExxonMobil being number one, and for hiring phony scientists. Just like in the tobacco cases.

I see a lot more project-specific litigation, such as against coal plants, the tar sands and pipelines, etc.

I think we're going to see a lot more litigation about the impacts of climate change such as the "Katrina Canal Breaches" case in which the U.S. Army Corps of Engineers was successfully sued by landowners in federal district court for damages because, the court upheld, the Corps had not properly maintained the regional flood protection system that was overwhelmed by Hurricane Katrina. If the case is affirmed on appeal, then there are several thousand or tens of thousands more lawsuits coming in which the damages could be in the tens of billions of dollars, payable by the Corps of Engineers. I see that as a foretaste of litigation against architects, engineers, and infrastructure providers who are going to be accused of not paying proper attention to predictions of sea-level rise and erratic weather and you didn't design against it, I was damaged and therefore you pay.[82]

That is a gargantuan amount of potential legal liability to which public authorities, insurance companies, design and engineering firms, and others need to pay heed. In the United States, Europe, Australia, and elsewhere, the law is becoming a critically important mover on climate and energy.

Being Aware

People are nearly everywhere becoming more aware of and better informed about climate, energy, and sustainability. This savvy is becoming baked into our educational systems, our laws, and our cultural institutions, and will only further deepen the understanding that people have of the issues and the many ways of dealing with the environmental crisis that confronts us.

This awareness is reflected in many polls. Most populations in the major economies that account for 80 percent of greenhouse gas emissions have a high level of awareness.[83] Gallup has surveyed people in 111 countries and found that 42 percent found global warming a serious threat.[84] In another poll, a broad majority wanted action on climate change, with majorities in fourteen of fifteen countries, both developed and developing nations, willing to address the matter.[85]

One other poll has found that about half of Americans recognize that particular phenomena such as coastline erosion and flooding, droughts, hurricanes, river flooding, and wildfires are being exacerbated by warming.[86]

Another study has found that, among Americans, there are six pronounced different levels of concern, from the "alarmed" to the "dismissive." The alarmed category was 10 percent when the survey was last done, with the "concerned" at 29 percent and the "cautious" at 27 percent. This third group believes, nevertheless, that warming is a problem, just not an urgent one. Two-thirds of Americans, according to the survey, therefore accept that there is a problem. The alarmed category is particularly strong and constitutes what political scientists call an "issue public" — a group that is highly engaged and motivated to address the issue.[87]

Among these concerned and active groups and individuals are some that one might not immediately associate with the issue. Hunters and anglers, for instance, a traditionally conservative bunch, have nevertheless shown real interest in climate change and have supported measures to reduce greenhouse gases. One such group, the National Wildlife Federation, founded in 1935 and with more than four million members and partners, has been active in mobilizing its constituencies to seek climate legislation in Congress.[88]

Faith Communities

Another large segment of the American population that is known for generally conservative views are those in various faith communities. The National Religious Partnership for the Environment, founded in 1993, however, is itself deeply concerned about climate change, as are its four main partner organizations. The partnership declares that "both air pollution and climate change are environmental justice issues because the poor and vulnerable are the least able to protect themselves from their effects."[89] One of the partners, the Coalition on the Environment and Jewish Life, "mobilizes the Jewish community to advocate on a wide range of environmental issues, with a particular focus on global climate change and energy conservation."[90] Another, the U.S. Conference of Catholic Bishops, is a key member of the Catholic Coalition on Climate Change, which has issued a "call for a civil dialogue and prudent and constructive action to protect God's precious gift of the earth's atmosphere."[91] Pope Benedict himself called "for strengthening the linkage between combating climate change and overcoming poverty."[92] The National Council of Churches of Christ says "climate change is a threat to all people and all of creation."[93] And one of the leaders of the Evangelical Environmental Network, Rev. Jim Ball, has a book, *Global Warming and the Risen LORD: Christian Discipleship and Climate Change*, in which he highlights "the biblical and spiritual resources we have been given to meet this threat."[94]

So the American environmental movement and faith communities are aligned in some very important ways. Two-thirds of Americans report that they care about environmental concerns because we are part of God's creation — and half of Sierra Club members say they attend services at least once a month.[95] "Creation Care" is a concept increasingly embraced by religious adherents in the United States and elsewhere. Evangelicals & Scientists United to Protect Creation is an organization with religious signatories as prominent as Jim Ball and Richard Cizik and environmental leaders with as high a profile as James Hansen, director of NASA's Goddard Institute for Space Studies; Gus Speth, founder of the World Resources Institute and former administrator of the UNDP; and distinguished ecologist Edward O. Wilson.[96]

Other religious leaders recognize the seriousness of the climate crisis. The Dalai Lama has said "I do have some serious concerns as a result of learning from specialists [that] unless we pay sufficient attention and [adopt] sufficient method of protection . . . global warming is really, really very serious."[97] Tibet and the Himalayas, as we have seen, are increasingly living under the Damoclean sword of climate change impacts.

The Consumption Divide

Much of the concern generated by the faith communities, in the United States and elsewhere, is a consequence of the many disparities between the developed and developing worlds. As we have seen, the impacts of climate change are reaching much more deeply and thoroughly into the lives of the poor in the developing world. The "North-South divide" is one way that this disparity has been framed, not only in the context of climate change, but in any number of ways relative to development.

Certainly, the question of responsibility for the climate crisis has been a key aspect in the international negotiations since 1992 and the work of the UN Framework Convention on Climate Change, as in most of the other multilateral discussions. The basic principle of "common but differentiated responsibilities" to address climate change embeds the idea that the developed world powers have a greater burden for addressing the crisis. As we have seen in the acceptance of the Kyoto Protocol by all the world's developed economies — except, notably, the United States — this principle is an accepted one. Even the United States, though, with the advent of the Obama administration, has been pushing for more responsibility by the developed economies, primarily through mechanisms such as financing for mitigation and adaptation, as well as technology transfer, and some others.

A good bit of the problem, then, lies in the extraordinary differences in rates of consumption. The highly developed economies simply use much more energy, water, and the essential components of postindustrial societies. In many respects, we are far too profligate in our use. Developing economies, and particularly those at the far end of the spectrum among the Least Developed Countries, use far fewer resources per capita. One calculation has it that the developed world has consumption rates roughly thirty-two times that of the developing world. Jared Diamond, a professor of geography and highly acclaimed author, wrote, "If the whole developing world were suddenly to catch up, world [consumption] rates would increase elevenfold. It would be as if the world population ballooned to 72 billion people (retaining present consumption rates)."[98]

It would be difficult to suppose that world resources, strained as they are today with a world population of seven billion — and making the many bad choices that we have been making—could support those levels of consumption. Consider, while we are at it, the output of GHGs and the further destruction of ecosystems from industrial agriculture and fishing, overdevelopment, and conventional water and air pollution from such galloping consumption. Are we thus doomed? Diamond's answer: "No, we could have a stable outcome in which all countries converge on consumption rates considerably below the current highest levels." But then we would have to reduce our access to labor-saving devices (driven by electricity, for the most part), and infant mortality would go up, gains in nutrition worldwide would evaporate, and literacy rates would plummet. Right? Wrong. Diamond reminds us, if we had forgotten, that "living standards are not tightly coupled to consumption rates."[99]

As we have seen a number of times here, the war against our forests and agricultural lands, the degradation of our increasingly precious freshwater resources and the marine environment, the escalating threat of the "Sixth Great Extinction," and the threat of catastrophic changes to our climate system have all served to motivate people to respond and are thus driving innovations and focused counterattacks on the forces that are causing these environmental depredations.

One big thought balloon that is appearing in more and more minds is the idea that we can consume less—and be better off for it. This means, of course, at the same time, that the people of the developing nations must be given much greater access to the basic components of a healthy life: clean air and water, healthy food, constantly improving economic prospects, educational opportunities for all—girls and boys, women and men—and a real stake and equality in the life of their communities and their nations, as well as in the community of nations. Franklin Roosevelt's "Four Freedoms"—freedom of speech and expression, freedom of worship, freedom from want, and freedom from fear—can certainly be achieved without an overreliance on hyper-industrialized economies, fueled primarily by fossil fuels, industrial chemicals, deforestation, etc., with the attendant greenhouse gases—fifty billion tons of CO_2eq annually.[100] There are hundreds of examples beyond the many examples that we have seen here that illustrate that this can be done.

How We Choose to Live

Technology policy and a "price on carbon" have been cited repeatedly as key components of the new approach to resource use that it is incumbent upon us to develop. These are, without question, critical. But lifestyle choices are also said to be at the core of the newer world that many hope to realize.

As we have seen, big retailers understand that more and more consumers are looking for products that they can feel good about, that have been produced sustainably, or that will make their homes and their lives less greenhouse gas intensive. The buy-in on hybrid vehicles has been large and continues to grow. Electric vehicles, in their infancy, are getting off to a healthy start. Mass transit is continuing to be a key component of urban planning all over the world. But bike paths, bike-sharing programs, and more pedestrian-friendly cities are also proliferating. In Germany, they are even rolling out a "car-sharing" program to enable intercity rail travelers to take advantage of electric car rentals.[101] Developers, renters, and homeowners have been clamoring for green buildings and more energy-efficient appliances. Compact fluorescent lightbulbs and light-emitting diodes (LEDs) are rapidly becoming the norm.

AC

Air conditioning has become ubiquitous, at least in America, and increasingly across the world. Ninety percent of new homes built in the United States have central air. The United States uses as much electricity for AC as the continent of Africa uses for all purposes. Period.

There are costs beyond exacerbating the problem of energy consumption and greenhouse gases. With AC, people stay inside more, exercise less, and get sick more often. The rightward tilt in American politics in the last fifty years may, according to one intriguing analysis, even be attributable to air conditioning: More people moved to the South and Southwest as they were better able to tolerate the heat, thus expanding the voting blocs that have altered Congress and the country. More people, faster economic growth, more congressional districts, and more Electoral College votes for these primarily conservative regions have meant more Republican Party influence on the national scene. That is the theory anyway.[102] In any event, it is, unfortunately, axiomatic that as societies become more affluent, their AC use skyrockets.

But as people are becoming more aware of their energy use and homes and offices become more efficient, the drain on energy from air conditioning should moderate. Most of the cooling that is actually going on never even reaches people. It cools the building itself. One of the many design innovations in the Bank of America Tower in New York City, designated LEED Platinum by the U.S. Green Building Council, is floor-level cooling, both in the enormous atrium in the lobby—so virtually no AC is dissipated into empty space—and in the individual offices, making it easier to control temperatures to the occupant's liking. The AC is generated by ice that has been made at night, during off-peak hours, saving money and strain on the grid. To top it all off, the BofA Tower also has a green roof, further insulating this already hugely efficient building.

Another burgeoning green building breakthrough is with the installation of many hundreds of thousands of geothermal (or ground-source) heat pumps all over the world in new buildings and as retrofits. These provide cooling as well as heating. This is another way of obviating the need for massive demand on electricity, particularly during the peak hours and days when it gets really hot. The number of units installed worldwide is expected to nearly double from almost three million in 2010 to 5.66 million in 2015.[103]

Moderating Energy Consumption

Electricity providers work with their customers to moderate energy use, very much including AC, when demand gets particularly heavy. As the smart grid becomes more of a reality, these demand-side management programs will increasingly be done automatically, because the "smart" appliances, like the AC unit in the home or business, will be programmed to respond.

One of the most salient and interesting phenomena regarding energy use is that when consumers are made aware of their own energy use relative to what their more energy-frugal neighbors are using, they are more apt to shave their own consumption with efficiency fixes and conservation.* "Home Energy Reports" contain this sort of information, which is much more than

---

* Good axiom: *Efficiency* is switching from an incandescent lightbulb to a CFL or LED. *Conservation* is turning off the light when you're not using it. From Ingrid Kelley, *Energy in America: A Tour of Our Fossil Fuel Culture and Beyond.*

the standard monthly utility bill. In a major study, data was analyzed from 750,000 households and found that these reports drive an average 1.8 percent reduction in household energy use. This modest reduction could translate, applied to all households in the United States, into a lowering of CO_2 emissions by almost nine million tons—the output of three 500 MW coal-fired power plants—and cost savings of $3 billion to consumers.[104] The study used data from OPOWER, a company that has "successfully converted large-scale customer engagement into a highly reliable energy efficiency program that delivers unprecedented energy savings to our utility partners."[105]

Japan, in the wake of the nuclear power catastrophe at Fukushima in March 2011, lost the power from those six reactors, plus they also suspended operations at another particularly vulnerable facility. With the advent of summer, the federal government asked citizens to cut back on their energy use. In particular, it re-emphasized the "Cool Biz" campaign initiated in 2005, asking businessmen and women to dress more lightly when it is hotter. The acceptance by the Japanese public of warmer office temperatures and cooler, but appropriate, business attire has been growing year by year. Japanese businessmen, like their counterparts in Hawaii, have been embracing the "kariyushi" shirt, a close cousin of the aloha shirt.[106]

These sorts of messages are increasingly pervasive. The U.S. Postal Service even has a "Go Green" program, including wonderful stamps that encourage environmentally sound behavior including composting, walking, biking, using public transportation, and adjusting the thermostat.[107]

Sustainable Tourism

People are not only concerned about their environmental footprint at home. They are taking their awareness on the road, too. This is particularly true when they are on vacation. Tourism provides more than 235 million jobs, with more than 9 percent of global GDP generated from the industry.[108] "Sustainable tourism" is, as you would guess, another rapidly growing area, because people want their experience to be environmentally sound.

Some of the major tourist destinations have always been in natural areas, from America's well-loved national and state parks to the coral reefs of the Pacific, from the mountains of Europe and North America where people can ski, hike, and camp, to the plains of Africa to see wildlife. The International Ecotourism Society (TIES) defines ecotourism as "responsible travel to natural areas that conserves the environment and improves the well-being of local people." TIES has more than five hundred members in over a hundred countries, works to promote the industry through conferences and other vehicles, offers training in sustainable tourism, and helped to inspire the International Year of Ecotourism declared by the UN for 2002.[109]

The Rainforest Alliance, UNEP, the United Nations Foundation, and the UN World Tourism Organization initiated the Global Sustainable Tourism Criteria in 2008. These are a set of thirty-seven voluntary standards provided to guide tourist destinations in how to sustain natural and cultural resources, enhance and increase tourist business, and alleviate poverty.[110]

UNEP coordinates the activities of the Global Partnership for Sustainable Tourism. The partnership has scores of members: national governments, UN organizations, the Organisation for Economic Co-operation and Development international business groups, and NGOs. Its mission is simply to "integrate sustainability into tourism."[111] An International Task Force on Sustainable Tourism Development, led by UNEP and France, produced a comprehensive set of policy recommendations with its report in 2009, the overall aims of which help stakeholders identify ways to improve business while guaranteeing sustainability and to promote awareness of and action on poverty alleviation and climate change.[112]

Voluntary Offsets:
Expanding the Possibilities for "Carbon Neutrality"

One of the ways that individuals and businesses have chosen to confront climate change is through the purchase of voluntary credits to "offset" their greenhouse gas footprint. You may, for instance, choose to make a donation to compensate for the greenhouse gas output of your air travel. The money you voluntarily contribute goes through a fund or other mechanism to validated offset projects, like a wind farm being built in South Korea, a landfill gas mitigation initiative in Colombia, or a project in Cambodia to preserve rainforest. Businesses also choose to buy offset credits in order to bank them with an eye to the future when they are likely to be regulated under a "compliance" regime. Businesses may also purchase offset credits in order to conform to standards of corporate social responsibility that they may have set for themselves.[113] The credits are known as Verified (or Voluntary) Emission Reductions (VERs) and are certified by a number of different organizations to ensure the purchase of "emission reductions and credits that are real, measurable, additional, permanent, conservatively estimated, independently verified, uniquely numbered and transparently listed in a central database."[114]

Of course, as we touched on in Chapter 5, these credits add up and are traded on the carbon markets. There were voluntary credits amounting to 131 million tons of CO_2eq traded in 2010, with a value of \$424 million. The volume and value are, to be sure, minuscule in comparison to that of the compliance markets — 6,692 $MtCO_2eq$ worth \$123.95 billion — but the voluntary markets have a certain "nonmarket" value in terms of providing funds for innovative projects and an outlet for ethical behavior for individuals and businesses.[115] These enterprises and the complex system of finance that support them have been in the vanguard of international environmental project development.

They also provide tremendous educational and practical value in helping people manage their greenhouse gases. The motto of one of the funds providing offsets is "Reduce what you can. Offset what you can't." Its suggestions as to how to reduce are practical and easy to do, with ideas applicable to the home, office, local and long-distance travel, your diet, and even your wedding![116]

Basics: Transport and Food

One of the most personal of preferences is how we get ourselves around. Bike sharing is growing exponentially in popularity in cities all over the world. There are as many as four hundred bike-sharing programs in just ten European countries. Paris and Barcelona were the grandparents of the movement, and the idea has spread to North America, East Asia, and Israel, among other places. New York City is gearing up to have ten thousand bicycles available at five hundred stations by the spring of 2012. Hangzhou, China, has the world's largest program, with 60,600 bikes and more than 2,400 stations.[117]

Of course, what we eat is another highly personal choice, driven by culture in most cases, that has a tremendous impact on the environment, not the least aspect being the climate system. The Dutch study referenced in the last chapter said that we could cut the cost in half of reaching the desirable goal of 450 ppm CO_2eq in the atmosphere by 2050 with a low-meat diet in place worldwide. The Indians put us all to shame with their per capita annual meat consumption of about five kilograms (11 pounds) per person. They have been a largely vegetarian society for centuries. The British, American, and French levels have remained relatively stable over the past twenty years, but the Germans reduced their consumption about 15 percent from 1990 to 2002. The Chinese, as with most things, have radically increased their levels of consumption. They doubled their meat consumption per capita to 52.4 kilograms (115.5 pounds) by 2002, and that figure has gone much higher again ten years later.[118]

No less a personage than Lord Nicholas Stern, the world's leading environmental economist working on climate change, recognizes the facts and calls for change. Perhaps even more important, he acknowledges the potential for lifestyle changes to occur:

> Meat is a wasteful use of water and creates a lot of greenhouse gases. It puts enormous pressure on the world's resources. A vegetarian diet is better. I think it's important that people think about what they are doing and that includes what they are eating. I am 61 now and attitudes towards drinking and driving have changed radically since I was a student. People change their notion of what is responsible. They will increasingly ask about the carbon content of their food.[119]

The Pace of Change

Not only for food choices but also for the whole kaleidoscope of other lifestyle changes, any number of expert commentators with long experience of working on environmental issues have the same perception as Lord Stern: that change can and does occur. As we saw in Chapter 7, Lester Brown's PBS documentary reminds us that "change which seems impossible happens."[120]

Can we effect the "convergence" of consumption that Jared Diamond and others recognize as necessary for the well-being of the planet's climate system? Will rapidly emerging economies like China and India be able to "leapfrog" from the development path that the major developed economies have taken with a near-total reliance on fossil fuels toward the much more environmentally benign "technology-driven" path that Hermann Scheer prescribed? Indeed, can the big economies of the G7 like the United States, Japan, Germany, and the UK make this transition?

In order to do this, we will need increasingly more aware and concerned publics driving political and private-sector leaders to make the right choices. These publics themselves have incredible power, both in how they spend their money and use resources, and in the messages they deliver at the voting booths and in other forums, such as in civil society vehicles they use to voice their concerns.

Indications are that we have been getting the message on climate change and sustainability, and that forces, perhaps inexorable, are moving to restore the balance and health to our planet and its ecosystems that are a reflection of its natural state and that are our birthright. It is certainly not a done deal and will require, if we are going to be successful, a continued and intensifying focus on making positive change happen, in our neighborhoods and personal lives, in our cities and our nations, and in the global village in which we are all now citizens.

Perhaps let Lord Tennyson's "Ulysses" inform humanity's present quest for sustainability:

> that which we are, we are;
> One equal temper of heroic hearts,
> Made weak by time and fate, but strong in will
> To strive, to seek, to find, and not to yield.

acknowledgments

Thanks to Davis Perkins and Bill Parkhurst, two old publishing pros who were good enough to read my proposal for this book and to give their approval.

To Greg Julian and Vera Jelinek, I am extremely grateful for being given the privilege of teaching. I also need to thank all my students for making me think more and study harder.

The Foreign Policy Association is to be thanked for giving me a platform for writing on climate change, energy, and sustainability.

Thanks also to Ron Puddu for his continuing support and encouragement.

I am grateful to the dedicated staff of the University Press of New England, including my editor, Stephen P. Hull, and to freelance copy editor Glenn Novak for his help in banging out the rough spots during copyediting.

Thanks to Lorraine Simonello for helping me with three figures. Thanks also to the staff of the various institutions and agencies that provided graphics and the requisite permissions.

I am particularly indebted to Dr. Gordon Hamilton of the Climate Change Institute of the University of Maine for his thorough and perceptive editorial input at two key points in the manuscript's preparation. I am also obliged to Marian Helms Hewitt, my in-house editor.

To those who graciously consented to be interviewed for the book and who provided such rich material, I am deeply grateful.

From the scientific community, Dr. Michael Oppenheimer of Princeton University gave generously of his time and expertise, as did Dr. Gavin Schmidt of NASA's Goddard Institute for Space Studies. Dr. Richard Houghton of Woods Hole Research Center was also helpful.

I got great insight from the journalists who cover energy, climate, and sustainability. These include Elizabeth Kolbert of the *New Yorker*, Elisabeth Rosenthal of the *New York Times*, and Vijay V. Vaitheeswaran of the *Econo-*

mist. Bill McKibben is another leading writer, who has also become a world-class activist, to whom I owe thanks. Jesse Berst, of the Center for Smart Energy, a leading writer and business consultant, was helpful.

From the nonprofit community, Tensie Whelan and Jeff Hayward of the Rainforest Alliance were extremely helpful, as was Mindy Lubber of Ceres. Mary Anne Hitt of the Sierra Club gave me some excellent material. Anna Lappé of the Small Planet Institute also provided valuable input.

From the business community, Todd Arnold of Duke Energy was also very helpful. Yvo de Boer of KPMG (and formerly of the UN Framework Convention on Climate Change) gave me some great interview material, as did Bob Fox of Cook + Fox Architects.

From academia, Dan Reicher of the Steyer-Taylor Center for Energy Policy and Finance at Stanford University (and formerly of Google), Dr. Nevin Cohen of the New School, and Professor Michael B. Gerrard of the Center for Climate Change Law at Columbia University all deserve my thanks.

I want to especially thank William K. Reilly for not only affording me the benefit of his unique insights across a range of salient subject matter but for being so kind as to provide this book with its eloquent foreword.

These people represent the broad community of writers, scientists, activists, public officials, entrepreneurs and business leaders, engineers, attorneys, designers, artists, and teachers who are so instrumental in piloting us toward this newer world to which we are bound.

notes

Introduction: Turning the Corner on the Climate Crisis

1 World Meteorological Organization press release, November 21, 2011.
2 Alex Steffen, "Geoengineering Megaprojects Are Bad Planetary Management," Worldchanging, February 9, 2009.

1. Science, Media, and the Public: The Message of Climate Change

1 "Climate Emails Stoke Debate," *Wall Street Journal*, November 23, 2009.
2 "E-Mail Fracas Shows Peril of Trying to Spin Science," *New York Times*, November 30, 2009.
3 Transcript of Glenn Beck from November 23, 2009, from *Watts Up with That?* blog.
4 "Climate Change: Chinese Adviser Calls for Open Mind on Causes," *Guardian*, January 24, 2010.
5 IPCC, Statement on the Melting of Himalayan Glaciers, January 20, 2010, http://www.ipcc.ch/pdf/presentations/himalaya-statement-20january2010 .pdf.
6 Report of the International Panel set up by the University of East Anglia to examine the research of the Climatic Research Unit, April 14, 2010, http:// www.uea.ac.uk/mac/comm/media/press/CRUstatements/SAP.
7 Netherlands Environmental Assessment Agency, "Assessing an IPCC Assessment: An Analysis of Statements on Projected Regional Impacts in the 2007 Report," July 5, 2010.
8 InterAcademy Council, "Climate Change Assessments: Review of the Processes and Procedures of the IPCC," August 30, 2010.
9 Peter Stott, global-average temperature records, Met Office website, http://www.metoffice.gov.uk/climate-change/guide/science/explained /temp-records.

10 James Hansen et al., Geophysical Research Letters, "Potential Climate Impact of Mount Pinatubo Eruption," August 17, 1991.

11 Worldwatch Institute, "James Hansen Talks about Climate Change," June 2008, http://www.worldwatch.org/node/5775.

12 Philip Shabecoff, "Global Warming Has Begun, Expert Tells Senate," *New York Times*, June 24, 1988.

13 Ad Hoc Study Group on Carbon Dioxide and Climate, Woods Hole, Massachusetts, July 23–27, 1979, to the Climate Research Board, Assembly of Mathematical and Physical Sciences, National Research Council, p. 8.

14 Michael Oppenheimer, interview with author, June 8, 2010.

15 Naomi Oreskes, testimony before the U.S. Senate Committee on Environment and Public Works, December 6, 2006.

16 Lyndon B. Johnson, "Special Message to the Congress on Conservation and Restoration of Natural Beauty," February 8, 1965.

17 Naomi Oreskes and Erik M. Conway, *Merchants of Doubt: How a Handful of Scientists Obscured the Truth on Issues from Tobacco Smoke to Global Warming* (New York: Bloomsbury Press, 2010), 177.

18 Ibid., 180.

19 Ibid., 181.

20 SCOPE 29—The Greenhouse Effect, Climatic Change, and Ecosystems, http://www.scopenvironment.org/downloadpubs/scope29/index.html.

21 Statement by the UNEP/WMO/ICSU International Conference, October 1985, http://www.scopenvironment.org/downloadpubs/scope29/statement.html.

22 *Report of the World Commission on Environment and Development: Our Common Future*, March 1987, http://www.un-documents.net/wced-ocf.htm.

23 IPCC First Assessment Report 1990 (FAR), Working Group 1—The IPCC Scientific Assessment, Policymakers Summary, p. xi.

24 William K. Reilly, "The Road from Rio: The Success of the Earth Summit Depends on How Well We Follow Through on Its Principles and Programs," *EPA Journal*, September/October 1992.

25 UN.org, UNCED information page.

26 Philip Shabecoff, *A Fierce Green Fire: The American Environmental Movement* (Washington, DC: Island Press, 2003), 192.

27 Oppenheimer interview.

28 Bill McKibben, e-mail exchange with author, October 28, 2010.

29 Maxwell T. Boykoff and J. Timmons Roberts, "Media Coverage of Climate

Change: Current Trends, Strengths, Weaknesses," United Nations Development Program, 2007, p. 36.

30 "Extreme Weather and Climate Change," Climate Central, November 16, 2011, http://www.climatecentral.org/features/extreme-weather-of-2011/.

31 Elizabeth Kolbert, interview with author, June 25, 2010.

32 Al Gore, *Our Choice: A Plan to Solve the Climate Crisis* (Emmaus, PA: Rodale Press, 2009), 18.

33 WorldPublicOpinion.org, "Multi-Country Poll Reveals That Majority of People Want Action on Climate Change, Even If It Entails Costs," December 3, 2009.

34 Jon A. Krosnick, "The Climate Majority," *New York Times* op-ed, June 8, 2010.

35 Yale Project on Climate Change Communication, "Poll: American Opinion on Climate Change Warms Up," June 10, 2010, http://news.yale.edu/2010 /06/08/poll-american-opinion-climate-change-warms.

36 www.350.org.

37 www.earthhour.org.

38 Bill McKibben, e-mail exchange with author, October 28, 2010.

39 Internet World Stats, "The Internet Big Picture," http://www.internetworld stats.com/stats.htm.

40 Naomi Oreskes, "Beyond the Ivory Tower: The Scientific Consensus on Climate Change," *Science*, December 3, 2004.

41 James Hoggan, *Climate Cover Up: The Crusade to Deny Global Warming* (Vancouver: Greystone, 2009).

42 Ross Gelbspan, *The Heat Is On: The Climate Crisis, the Cover-up, the Prescription* (Boston: Addison-Wesley, 1997).

43 Hoggan, *Climate Cover Up*, 1.

44 Gelbspan, *Heat Is On*, 5.

45 Greenpeace, "Dealing in Doubt: The Climate Denial Industry and Climate Science," March 2010, p. 2, http://www.greenpeace.org/international/Global /international/planet-2/report/2010/3/dealing-in-doubt.pdf.

46 Interview from "Hot Politics," *Frontline*, November 13, 2006.

47 Michael Oppenheimer, *Nightline*, ABC News, February 24, 1994.

48 Gavin Schmidt and Joshua Wolfe, eds., *Climate Change: Picturing the Science* (New York: W. W. Norton, 2009), 7.

49 Aaron M. McCright and Riley E. Dunlap, "Defeating Kyoto: The Conservative Movement's Impact on U.S. Climate Change Policy," *Social Problems* 50, no. 3 (August 2003): 348–73.

50 "A Reporter's Field Notes on the Coverage of Climate Change," *Yale Environment 360*, March 11, 2009.

51 Kolbert interview.

52 Ibid.

53 Climate Central, "History," http://www.climatecentral.org/what-we-do /history/.

54 Gavin Schmidt, interview with author, June 21, 2010.

55 Elizabeth Kolbert, "Reporting on Climate Change," in Schmidt and Wolfe, *Climate Change*, 71.

56 John Broder, "Scientists Taking Steps to Defend Work on Climate," *New York Times*, March 2, 2010.

57 William Scott, letter to the editor, *New York Times*, March 4, 2010.

58 Matthew C. Nisbet, "Climate Shift — Clear Vision for the Next Decade of Public Debate," ClimateShiftProject.org, p. 57.

59 Deutsche Welle, "The Heat Is On — Climate Change and the Media," November 2010.

60 "Anthony Leiserowitz on Global Warming's 'Six Americas,'" Yale Forum on Climate Change and the Media, October 23, 2010, http://environment.yale .edu/climate/multimedia/anthony-leiserowitz-on-global-warmings-six -americas/.

61 Jeff Tollefson, "Educators Take Aim at Climate Change," *Nature*, October 26, 2010.

62 Elisabeth Rosenthal, interview with author, October 28, 2010.

63 Schmidt interview.

64 Simon Shuster, "Will Russia's Heat Wave End Its Global-Warming Doubts?" *Time*, August 2, 2010.

65 David Remnick, "Ozone Man," *New Yorker*, April 24, 2006.

66 "Global Consumers Vote Al Gore, Oprah Winfrey and Kofi Annan Most Influential to Champion Global Warming Cause," Nielsen Survey, July 2, 2007, http://nz.nielsen.com/news/GlobalWarming_Jul07.shtml.

67 Ole Danbolt Mjøs, Nobel presentation speech, December 10, 2007.

68 Al Gore, Nobel lecture, December 10, 2007.

69 Communication with Rodale Press publicity office, November 5, 2010.

2. Green Power: Renewable Energy Comes into Its Own

1 Carbon Dioxide Information Analysis Center, U.S. Department of Energy, http://cdiac.ornl.gov/trends/emis/tre_glob.html.

2 Correspondence by author with Dr. Richard Houghton, deputy director and senior scientist, Woods Hole (MA) Research Center, June 30, 2010.

3 "The Pre-industrial Carbon Dioxide Level," Climatic Change, May 5, 2006, http://www.springerlink.com/content/a06440223792h163/.

4 NOAA Earth System Research Laboratory, "Trends in Atmospheric Carbon Dioxide," http://www.esrl.noaa.gov/gmd/ccgg/trends/.

5 World Resources Institute, "Global Emissions of CO_2 from Fossil Fuels: 1900–2004," http://www.wri.org/chart/global-emissions-co2-from-fossil-fuels-1900-2004.

6 Jeff Goodell, *Big Coal: The Dirty Secret behind America's Energy Future* (New York: Houghton Mifflin Harcourt, 2006), xiii.

7 World Coal Association, http://www.worldcoal.org/coal/uses-of-coal/coal-electricity/.

8 Joe Romm, "Coal Power Drops below 40% of U.S. Electricity, Lowest in 33 Years," Climate Progress, March 12, 2012.

9 Michelle L Bell, Devra L Davis, and Tony Fletcher, "A Retrospective Assessment of Mortality from the London Smog Episode of 1952: The Role of Influenza and Pollution," *Environmental Health Perspectives* 112, no. 1 (January 2004): 6–8, http://www.ncbi.nlm.nih.gov/pmc/articles/PMC1241789/.

10 Alan H. Lockwood, Kristen Welker-Hood, Molly Rauch, and Barbara Gottlieb, "Coal's Assault on Human Health," Physicians for Social Responsibility, November 2009, http://www.psr.org/resources/coals-assault-on-human-health.html.

11 EPA Office of Air Quality Planning & Standards and Office of Research and Development. *Mercury Study Report to Congress*, vol. 2, *An Inventory of Anthropogenic Mercury Emissions in the United States*, December 1997, EPA-452/R-97-004.

12 Richard McGregor, "750,000 a Year Killed by Chinese Pollution," *Financial Times*, July 2, 2007.

13 Centers for Disease Control and Prevention, http://www.cdc.gov/niosh/mining/statistics/images/pp3.gif.

14 Shai Oster and Gordon Fairclough, "China to Crack Down on Mine Safety Violations," *Wall Street Journal*, April 7, 2010.

15 Elizabeth Shogren, "Tennessee Spill: The Exxon Valdez of Coal Ash?" NPR, December 31, 2008.

16 John McQuaid, "Mining the Mountains," *Smithsonian*, January 2009.

17 Jeffrey D. Sachs and Andrew M. Warner, "Natural Resource Abundance and Economic Growth," Center for International Development and Harvard Institute for International Development, November 1997.

18 Goodell, *Big Coal*, 32.

19 Hermann Scheer, ACORE speech, November 20, 2009, http://www.c-spanvideo .org/program/id/215782 (from 45:50).

20 IHS CERA, press release, February 5, 2008.

21 Dan Reicher, interview with author, July 9, 2010.

22 U.S. Energy Information Administration, International Electricity Installed Capacity, http://www.eia.gov/emeu/international/electricitycapacity.html.

23 U.S. Energy Information Administration, International Energy Statistics, http://www.eia.gov/cfapps/ipdbproject/IEDIndex3.cfm?tid=44&pid =44&aid=1.

24 Exajoule — 10^{18} joules. A joule is the work required to continuously produce 1 watt of power for one second.

25 Thomas B. Johansson, Kes McCormick, Lena Neij, and Wim Turkenburg, "The Potentials of Renewable Energy, Thematic Background Paper," International Conference for Renewable Energies, Bonn 2004, http://www .renewables2004.de/doc/DocCenter/TBP10-potentials.pdf.

26 "Renewable Energy Snapshots 2010," European Commission Joint Research Centre, Institute for Energy, July 2010, http://re.jrc.ec.europa.eu/refsys/pdf /FINAL_SNAPSHOTS_EUR_2010.pdf.

27 "Wind in Power — 2011 European Statistics," European Wind Energy Association, February 2012, http://ewea.org/fileadmin/ewea_documents /documents/publications/statistics/Stats_2011.pdf.

28 "The Greenest Bail-out?" *Financial Times*, February 2, 2009.

29 REN21, Renewables 2010 Global Status Report, July 15, 2010, http://www .ren21.net/Portals/97/documents/GSR/REN21_GSR_2010_full_revised %20Sept2010.pdf.

30 "India's Renewable Energy Generation Capacity Could Reach 48 GW by 2015," CleanTechnica.com, July 25, 2010.

31 Greenpeace and the European Renewable Energy Council (EREC), "EU Energy [R]evolution," June 2010, http://www.greenpeace.org/eu-unit/Global /eu-unit/reports-briefings/2010/7/EU-Energy-%28R%29-evolution-scenario.pdf.

32 "Energy Goal for 2050: 100% Renewable Electricity Supply," Federal Environment Agency, press release, July 7, 2010.

33 Centre for Alternative Technology, ZeroCarbonBritain2030, http://www.zerocarbonbritain.org/.

34 Repower America, http://www.facebook.com/repoweramerica.

35 The White House, Weekly Address, July 3, 2010.

36 "How We Can Take Firm Action to Cut Carbon," *Financial Times*, January 24, 2008.

37 Global Wind Energy Council, "Global Wind Report 2010," http://www.gwec.net/fileadmin/images/Publications/GWEC_annual_market_update_2010_-_2nd_edition_April_2011.pdf.

38 Global Wind Energy Council, press release, January 29, 2010.

39 Crown Estate, press release, January 8, 2010.

40 "China Plans to Boost Efforts to Develop Wind-Power Generating Capacity," *Bloomberg News*, June 7, 2010.

41 U.S. Department of the Interior, press release, June 8, 2010.

42 Matthew Wald, "Offshore Wind Power Line Wins Backing," *New York Times*, October 12, 2010.

43 Department of the Interior, press release, February 7, 2011.

44 Oregon Power Solutions, http://www.oregonpowersolutions.org/index.php?option=com_content&task=view&id=15&Itemid=35.

45 "Hywind — the World's First Full-Scale Floating Wind Turbine," Statoil webpage, http://www.statoil.com/en/TechnologyInnovation/NewEnergy/RenewablePowerProduction/Offshore/Hywind/Pages/HywindPuttingWindPowerToTheTest.aspx.

46 Andrew Williams, "A Buoyant Future for Floating Wind Turbines?" *Renewable Energy World*, May 3, 2011.

47 Global Wind Energy Council, "The Visual Impact of Wind Turbines," http://www.gwec.net/index.php?id=141&L=0.

48 Greenpeace USA, "Bill Koch: The Dirty Money behind Cape Wind Opposition," July 22, 2010, http://www.greenpeace.org/usa/en/news-and-blogs/campaign-blog/bill-koch-the-dirty-money-behind-cape-wind-op/blog/26104/.

49 "Preparation for Groundbreaking Offshore Wind Farm Project Begins in Atlantic City," *Newark Star-Ledger*, May 1, 2010.

50 Lester Brown, *Plan B 4.0: Mobilizing to Save Civilization* (New York: W. W. Norton, 2009), 116.

51 Pike Research, "Executive Summary — Global Wind Energy Outlook," July

2011, http://www.pikeresearch.com/wordpress/wp-content/uploads/2011/07/WIND-11-Executive-Summary.pdf.

52 Global Wind Energy Council, "Global Wind Energy Outlook 2008," Investment and Employment, http://www.gwec.net/fileadmin/documents/GWEO08-graphs/GWEO_A4_2008_42b.jpg.

53 Brown, *Plan B 4.0*, xiii.

54 Manfred Körner, chairman, Umweltforum, letter in *Financial Times*, February 9, 2010.

55 "Direct Solar Energy," *Special Report on Renewable Energy Sources and Climate Change Mitigation* (SRREN), IPCC, May 2011, p. 9.

56 World Bank, *World Development Report 2010*, "Global Direct Normal Solar Radiation," http://siteresources.worldbank.org/INTWDR2010/Resources/5287678-1226014527953/WDR10-Full-Text.pdf.

57 IHS Emerging Energy Research, "Global Concentrated Solar Power Markets and Strategies: 2010–2025," April 2010.

58 Greenpeace International, SolarPACES and ESTELA, "Concentrating Solar Power Outlook 2009," May 2009, http://www.greenpeace.org/international/Global/international/planet-2/report/2009/5/concentrating-solar-power-2009.pdf.

59 Desertec Industrial Initiative, http://www.dii-eumena.com/home.html.

60 EurActiv.com, "EU Sees Solar Power Imported from Sahara in Five Years," June 23, 2010.

61 Author's notes from Climate Change and Global Security panel, American Museum of Natural History, June 25, 2010.

62 Abengoa, http://www.abengoasolar.com/corp/web/en/index.html/.

63 Amory B. Lovins, "Four Nuclear Myths," Rocky Mountain Institute, October 13, 2009, p. 5 (italics in original), http://www.rmi.org/Knowledge-Center/Library/2009-09_FourNuclearMyths.

64 Abengoa, http://www.abengoasolar.com/corp/web/en/index.html/.

65 Maria Gallucci, "Solyndra Shakeout Seen as a Sign of Success for Wider Solar Market," *InsideClimate News*, September 21, 2011.

66 Lawrence Berkeley National Laboratory, press release, September 15, 2011.

67 BrightSource, http://www.brightsourceenergy.com/projects/ivanpah.

68 International Geothermal Association, http://www.geothermal-energy.org/index.php.

69 International Energy Agency, Technology Roadmap — Geothermal Heat and Power, June 14, 2011, p. 9, http://www.iea.org/papers/2011/Geothermal_Roadmap.pdf.

70 Leonora Walet and Tessa Dunlop, "Analysis: Can Geothermal Help Japan in Crisis?" Reuters, March 24, 2011.

71 Google, Enhanced Geothermal Systems webpage, FAQ, http://www.google.org/egs/faq.html.

72 Reicher interview.

73 "Unconventional Oil and Gas 'Hot Areas,'" *Financial Times*, March 8, 2010.

74 "Ocean Energy," *Special Report on Renewable Energy Sources and Climate Change Mitigation*, IPCC, May 9, 2011, p. 4.

75 Ibid.

76 Pike Research, press release, January 19, 2010.

77 Crown Estate, press release, March 16, 2010.

78 David Fogarty, "Ocean Waves Can Power Australia's Future, Scientists Say," Planet Ark, August 20, 2010.

79 "Tidal Power Primed for Breakthrough," International Water Power and Dam Construction, February 17, 2009, http://www.waterpowermagazine.com/story.asp?sc=2052179.

80 Kyunghee Park, "GS Engineering to Construct World's Largest Tidal Power Plant," *Bloomberg News*, January 19, 2010.

81 Power Plants around the World, http://www.industcards.com/top-100-pt-1.htm

82 U.S. Energy Information Administration, International Energy Statistics, http://www.eia.gov/cfapps/ipdbproject/iedindex3.cfm?tid=2&pid=alltypes&aid=12&cid=&syid=2004&eyid=2007&unit=BKWH.

83 "Hydropower," "Ocean Energy," *Special Report on Renewable Energy Sources and Climate Change Mitigation*, IPCC, May 9, 2011, p. 4.

84 "Energy Supply," chap. 4 in "Climate Change 2007: Mitigation; Contribution of Working Group III to the Fourth Assessment Report of the Intergovernmental Panel on Climate Change," pp. 273–74, http://www.ipcc.ch/pdf/assessment-report/ar4/wg3/ar4-wg3-chapter4.pdf.

85 International Rivers, "Three Gorges Dam," http://www.internationalrivers.org/china/three-gorges-dam.

86 "Industry, Settlement and Society," chap. 7 in "Climate Change 2007: Impacts, Adaptation and Vulnerability; Contribution of Working Group II to the Fourth Assessment Report of the Intergovernmental Panel on Climate Change," p. 367, http://www.ipcc.ch/pdf/assessment-report/ar4/wg2/ar4-wg2-chapter7.pdf.

87 National Hydropower Association, "Pumped Storage," http://hydro.org/tech-and-policy/technology/pumped-storage/.

88 "Bioenergy," *Special Report on Renewable Energy Sources and Climate Change Mitigation*, IPCC, May 9, 2011, p. 5.

89 Ibid.

90 Joseph Fargione et al., "Science, Land Clearing and the Biofuel Carbon Debt," *Science*, February 29, 2008, http://www.sciencemag.org/content/319 /5867/1235.abstract#aff-1.

91 David Pimentel et al., "Human Ecology, Food versus Biofuels: Environmental and Economic Costs," *Human Ecology*, January 29, 2009, http://www .springerlink.com/content/47705417208688m7/.

92 ACORE and EPRI, "Reinventing Renewable Energy," July 2009, p. 6, http://74.220.216.101/acore/wp-content/uploads/2011/02/ACORE-EPRI -FINAL11.pdf.

93 "Mali's Farmers Discover a Weed's Potential Power," *New York Times*, September 9, 2007.

94 "Biochar as the New Black Gold," Grist, August 12, 2009.

95 "Sustainable Biochar to Mitigate Global Climate Change," *Nature Communications*, July 14, 2010.

96 Department of Energy, press release, September 15, 2009.

97 "Venice Seaport Eyes Algae to Fuel Energy Needs," Reuters, March 25, 2009.

98 Synthetic Genomics Inc., press release, July 14, 2010.

99 Amory Lovins, from speech at Urban Green Expo, September 23, 2009.

100 Greenpeace International, *The Silent Energy [R]evolution: 20 Years in the Making*, June 2011, p. 4, http://www.greenpeace.org/international/en /publications/The-silent-Energy-Revolution/.

3. Bringing It All Back Home: Clean Tech in Action

1 U.S. Energy Information Administration, "Electricity Flow, 2010," http:// www.eia.gov/totalenergy/data/annual/pdf/sec8_3.pdf.

2 Nicholas Z. Muller et al., "Environmental Accounting for Pollution in the United States Economy," *American Economic Review* 101, no. 5 (August 2011).

3 Javier Blas, "IEA Counts $550bn Energy Support Bill," *Financial Times*, June 6, 2010.

4 Muriel Boselli, "IEA Warns of Ballooning World Fossil Fuel Subsidies," Reuters, October 4, 2011.

5 Environmental Law Institute, "Estimating U.S. Government Subsidies to

Energy Sources: 2002–2008," September 2009, http://www.elistore.org /reports_detail.asp?ID=11358.

6 OECD-IEA, "Fossil Fuel Subsidies and Other Support," http://www.oecd .org/site/0,3407,en_21571361_48776931_1_1_1_1_1,00.html.

7 Kari Manlove, "Energy Poverty 101," Center for American Progress, May 14, 2009.

8 World Alliance for Decentralized Energy (WADE), http://www.localpower .org/.

9 "U.S. Military Is Developing Smart Microgrids with Solar Power," Clean-Technica, June 18, 2010.

10 U.S. Department of Energy, Combined Heat and Power Partnership, http:// www.epa.gov/chp/index.html.

11 New York University, press release, January 21, 2011.

12 "Facebook Photos Could Warm Homes," CleanTechnica, June 27, 2010.

13 "Data Center under Helsinki to Warm Residents Above," CleanTechnica, March 14, 2010.

14 "Europe Finds Clean Energy in Trash, but U.S. Lags," *New York Times*, April 12, 2010.

15 Elizabeth Kolbert, "The Island in the Wind," *New Yorker*, July 7, 2008.

16 "Home-Grown Power," *New York Times*, March 6, 2009.

17 "Solar Heat Worldwide 2011," IEA Solar Heating and Cooling Programme, May 2011, http://www.iea-shc.org/publications/downloads/Solar_Heat _Worldwide-2011.pdf.

18 "Australian Solar Rides the Bull," RenewableEnergyWorld.com, July 2, 2010.

19 Ucilia Wang, "DOE Backs Largest Rooftop Solar Project in U.S.," *GigaOm*, June 22, 2011.

20 Jesse Morris, "Largest U.S. Solar Rooftop Project Secures $1.4 Billion DoE Loan Guarantee," Rocky Mountain Institute, June 30, 2011.

21 Department of Energy, press release, September 7, 2011.

22 Solar Energy Industries Association, press release, September 20, 2011.

23 The Solar Foundation, press release, September 19, 2011.

24 European Photovoltaic Industry Association (EPIA), press release, May 4, 2011.

25 U.S. Energy Information Administration, International Electricity Installed Capacity, http://www.eia.gov/emeu/international/electricitycapacity.html.

26 "Wind Powered Factories: History (and Future) of Industrial Windmills," *Low-tech Magazine*, October 8, 2009.

27 2010 American Wind Energy Association Small Wind Turbine Global Market Study, http://www.awea.org/learnabout/smallwind/upload/2010_AWEA_Small_Wind_Turbine_Global_Market_Study.pdf.

28 Pike Research, press release, September 30, 2011.

29 IEA/OECD Heat Pump Programme, "Heat Pumps Can Cut Global CO_2 Emissions by Nearly 8%," 2008, http://www-v2.sp.se/hpc/publ/HPCOrder/viewdocument.aspx?RapportId=451.

30 GeoExchange, "Fascinating Facts," http://www.geoexchange.org/downloads/GB-003.pdf.

31 U.S. Department of Energy, "How Fuel Cells Work," http://www.fueleconomy.gov/feg/fcv_PEM.shtml.

32 National Hydrogen Association, Hydrogen Production Overview, August 2004, http://www.fchea.org/core/import/PDFs/factsheets/Hydrogen%20Production%20Overview_NEW.pdf.

33 MIT, press release, July 31, 2008.

34 Bloom Energy, press release, February 24, 2010.

35 Daimler, press release, September 9, 2009.

36 Pike Research, press release, October 7, 2011.

37 European Hydrogen Association, "EU Fuel Cell Aircraft Makes Test Flights," July 16, 2010, http://www.h2euro.org/latest-news/hydrogen-hits-the-roads/eu-fuel-cell-aircraft-makes-test-flights.

38 Worldwatch Institute, oil consumption graph, http://www.worldwatch.org/brain/images/press/news/vs05-world_oil.jpg.

39 Daniel Yergin, *The Prize: The Epic Quest for Oil, Money, and Power* (New York: Free Press, 1993), 523.

40 World Resources Institute—Climate Analysis Indicators Tool, Figures, http://cait.wri.org/figures.php.

41 Vijay V. Vaitheeswaran, interview with author, May 19, 2010.

42 "A Future That Doesn't Guzzle," *New York Times*, January 11, 2010.

43 "Nissan: Electric Cars Could Shed Government Aid in 4 Years," Reuters, May 8, 2010.

44 Tesla, press release, July 16, 2010.

45 GM, press release, July 27, 2010.

46 GM, press release, May 18, 2011.

47 Energy Source, *Financial Times*, May 27, 2010.

48 Dan Reicher, interview with author, July 9, 2010.

49 Pike Research, press release, August 22, 2011.

50 U.S. Department of Energy, "The Recovery Act: Transforming America's Transportation Sector Batteries and Electric Vehicles," July 14, 2010.

51 Pike Research, press release, August 24, 2011.

52 "Better Place CEO Shai Agassi: With 1 Billion Cars by 2015, Electric Car Has Built-in Audience," SmartPlanet, July 16, 2010.

53 Vaitheeswaran interview.

54 "Variability versus Predictability of Wind Power Production," EWEA, http://www.wind-energy-the-facts.org/en/part-2-grid-integration/chapter-2-wind-power-variability-and-impacts-on-power-systems/variability-versus-predictability-of-wind-power-production.html.

55 Reicher interview.

56 National Hydropower Association, "Pumped Storage," http://hydro.org/tech-and-policy/technology/pumped-storage/.

57 Samir Succar and Robert H. Williams, "Compressed Air Energy Storage: Theory, Resources, and Applications for Wind Power," Princeton Environmental Institute, April 8, 2008.

58 "World's First Molten Salt Concentrating Solar Power Plant Opens," CleanTechnica, July 26, 2010.

59 U.S. Department of Energy, "Hawaiian Wind Farm Project Producing Jobs," July 23, 2010.

60 Ice Energy, press release, January 27, 2010.

61 The Durst Organization, fact sheet, http://www.durst.org/assets/pdf/data sheet/OneBryantPark.pdf.

62 Bob Fox, interview with author, June 22, 2010.

63 Fraunhofer-Gesellschaft, "Storing Green Electricity as Natural Gas," May 5, 2010.

64 University of Delaware, "Vehicle to Grid Technology," http://www.udel.edu/V2G/index.html.

65 Amory Lovins, "Four Nuclear Myths: A Commentary on Stewart Brand's Whole Earth Discipline and on Similar Writings," Rocky Mountain Institute, October 13, 2009, p. 8, http://www.rmi.org/Knowledge-Center/Library/2009-09_FourNuclearMyths.

66 Friends of the Supergrid, http://www.friendsofthesupergrid.eu/.

67 Pike Research, press release, June 30, 2010.

68 Pike Research, press release, July 21, 2011.

69 Todd Arnold, Duke Energy senior vice president for the smart grid, author's interview, September 2008.

70 "Understanding the Benefits of the Smart Grid," National Energy Technology Laboratory, June 18, 2010, p. 6, http://www.netl.doe.gov/smartgrid/referenceshelf/whitepapers/06.18.2010_Understanding%20Smart%20Grid%20Benefits.pdf.

71 William F. Hewitt, "Current Concerns," Planning, December 2008, http://foreignpolicyblogs.com/wp-content/uploads/current-concerns.pdf.

72 Pike Research, press release, June 30, 2011.

73 Pike Research, press release, March 7, 2011.

74 Pike Research, press release, September 16, 2011.

75 Al Gore, Our Choice: A Plan to Solve the Climate Crisis (Emmaus, PA: Rodale Press, 2009), 288.

76 Ibid.

77 "Utility Decoupling: Giving Utilities Incentives to Promote Energy Efficiency," Progressive States Network, September 10, 2007, http://www.progressivestates.org/blog/672/utility-decoupling-giving-utilities-incentives-to-promote-energy-efficiency.

78 Vaitheeswaran interview.

79 U.S. Department of Energy, "Successes of the Recovery Act," January 2012, http://energy.gov/sites/prod/files/RecoveryActSuccess_Jan2012final.pdf.

80 Jesse Berst, interview with author, September 2008.

81 William F. Hewitt, "The Green Building Movement," Foreign Policy Association, July 2007, http://www.fpa.org/topics_info2414/topics_info_show.htm?doc_id=511664.

82 Ibid.

83 "Buildings and Climate Change," UNEP-Sustainable Buildings and Climate Initiative (SBCI), April 2009, http://www.unep.org/sbci/pdfs/SBCI-BCC Summary.pdf.

84 "Pathways to a Low-Carbon Economy," McKinsey and Co., 2009, https://solutions.mckinsey.com/ClimateDesk/default.aspx.

85 Empire State Building, "Sustainability & Energy Efficiency," http://www.esbnyc.com/sustainability_energy_efficiency.asp.

86 "Green Building and Human Experience," US Green Building Council (USGBC), June 2010, http://www.usgbc.org/ShowFile.aspx?DocumentID=7383.

87 Fox interview.

88 Lawrence Berkeley National Laboratory, press release, July 19, 2010.

89 Green Jobs Study, USGBC / Booz Allen Hamilton, November 2009, http://www.usgbc.org/ShowFile.aspx?DocumentID=6435.

90 Climate Savers Computing Initiative, 2010 Progress Report, June 2010, http://www.climatesaverscomputing.org/docs/2010-Progress-Report.pdf.

91 "IT Aims to Save the World: The State of Green Business 2010," GreenBiz.com, February 10, 2010.

92 IBM's "Smarter Planet," http://www.ibm.com/smarterplanet/us/en/.

4. Breakthroughs: From Rio to Cancún, and Beyond

1 Summit on Climate Change, "Summary by the Secretary-General," September 22, 2009, http://www.un.org/wcm/webdav/site/climatechange/shared/Documents/Chair_summary_Finall_E.pdf.

2 UNFCCC, meeting archives, COP 15, http://unfccc.int/meetings/copenhagen_dec_2009/meeting/6295.php.

3 Jim Tankersley, "Climate Summit Ends with Major Questions: 'Breakthrough' or 'Cop-out'?" *Los Angeles Times*, December 19, 2009.

4 Greenpeace, press release, December 19, 2009.

5 Ed Miliband, "The Road from Copenhagen," *Guardian*, December 20, 2009.

6 Geoff Dyer, "Beijing Rejects UK Copenhagen Criticism," *Financial Times*, December 22, 2009.

7 Philip Shabecoff, *A Fierce Green Fire: The American Environmental Movement* (Washington, DC: Island Press, 2003), 111.

8 Nairobi Declaration, May 18, 1982, http://hqweb.unep.org/Law/PDF/NairobiDeclaration1982.pdf.

9 *Report of the World Commission on Environment and Development: Our Common Future*, March 1987, chap. 2, paragraph 15.

10 William K. Reilly, interview with author, December 10, 2010.

11 Michael E. Kraft, *Environmental Policy and Politics* (Longman, 2007), 2.

12 William K. Reilly, "The Road from Rio: The Success of the Earth Summit Depends on How Well We Follow Through on Its Principles and Programs," *EPA Journal*, September/October 1992.

13 Reilly interview.

14 Kraft, *Environmental Policy and Politics*, 2.

15 Text of the UN Framework Convention on Climate Change.

16 Ibid.

17 Byrd-Hagel Resolution, U.S. Senate, July 25, 1997.

18 Transcript of President George W. Bush's speech on global climate change,

June 11, 2001, at *Washington Post* website, http://www.washingtonpost.com/wp-srv/onpolitics/transcripts/bushglobal_061101.htm.

19 Climate Change 2001: Synthesis Report—Summary for Policymakers, Question 2.9-11, September 2001, http://www.grida.no/climate/ipcc_tar/vol4/english/pdf/spm.pdf.

20 USGCRP, *Climate Action Report 2002*, chap. 6, "Impacts and Adaptation," May 2002.

21 Kraft, *Environmental Policy and Politics*, 253.

22 Guus J. M. Velders et al., "The Importance of the Montreal Protocol in Protecting Climate," *PNAS* (*Proceedings of the National Academy of Sciences*) 104, no. 12 (March 20, 2007): 4814–19.

23 *Climate Change Reference Guide*, Worldwatch Institute, p. 4, http://www.worldwatch.org/files/pdf/CCRG.pdf.

24 Drew T. Shindell et al., "Improved Attribution of Climate Forcing to Emissions," *Science*, October 30, 2009, 716–18.

25 "Canada to Withdraw from Kyoto Protocol," BBC, December 13, 2011.

26 UNFCCC, Compliance under the Kyoto Protocol, http://unfccc.int/kyoto_protocol/compliance/items/2875.php.

27 UNFCCC, JI Projects, RU1000200: Yety-Purovskoe Oil Field Associated Gas Recovery and Utilization Project, http://ji.unfccc.int/JIITLProject/DB/230U9F5CNZ0FTHULB2OB6HQMXWJ0ER/details.

28 European Union, press release, October 7, 2011.

29 "EU Climate Package Explained," BBC, April 9, 2010.

30 Chris Huhne, Norbert Röttgen, and Jean-Louis Borloo, "Europe Needs to Reduce Emissions by 30%," *Financial Times*, July 14, 2010.

31 David Fogarty, "Nod for Australia's Labor Likely Boost for CO_2 Law," Reuters (Analysis), September 8, 2010.

32 Arthur Max, "Obama Buzz Felt at Global Climate Talks," Associated Press, November 29, 2008.

33 Brian Urquhart, "What You Can Learn from Reinhold Niebuhr," *New York Review of Books*, March 26, 2009.

34 David Adam, "Nicholas Stern: Spend Billions on Green Investments Now to Reverse Economic Downturn and Halt Climate Change," *Guardian*, February 11, 2009.

35 Council of Economic Advisers, *The Economic Impact of the American Recovery and Reinvestment Act of 2009: Fourth Quarterly Report, July 14, 2010*, table 12.

36 "President Obama Announces National Fuel Efficiency Policy," White House press release, May 19, 2009.

37 "President Obama Directs Administration to Create First-Ever National Efficiency and Emissions Standards for Medium- and Heavy-Duty Trucks," White House press release, May 21, 2010.

38 "Remarks by the President on Energy," White House, June 29, 2009.

39 "President Obama Signs an Executive Order Focused on Federal Leadership in Environmental, Energy, and Economic Performance," White House, October 5, 2009.

40 "US Air Force: We Want to Use Biofuels," Agence France-Presse, July 19, 2011.

41 Elisabeth Rosenthal, "U.S. Military Orders Less Dependence on Fossil Fuels," *New York Times*, October 4, 2010.

42 Thomas Friedman, "The U.S.S. Prius," *New York Times*, December 18, 2010.

43 Pike Research, press release, October 13, 2011.

44 U.S. Global Change Research Program, *Our Changing Planet: The U.S. Global Change Research Program for Fiscal Year 2011*, p. 68, http://downloads .globalchange.gov/ocp/ocp2011/ocp2011.pdf.

45 Department of the Interior, press release, "Salazar Signs First U.S. Offshore Commercial Wind Energy Lease with Cape Wind Associates, LLC," October 6, 2010.

46 EPA — "EPA's Endangerment Finding — Legal Background," December 7, 2009.

47 "President Obama Announces Launch of the Major Economies Forum on Energy and Climate," White House press statement, March 28, 2009.

48 Major Economies Forum on Energy and Climate, fact sheet, April 2010.

49 The G-20, *The Pittsburgh Summit: Acting on Our Global Energy and Climate Change Challenges*, September 2009, http://www.whitehouse.gov/files /documents/g20/Pittsburgh_Fact_Sheet_Energy_Security.pdf.

50 Javier Blas, "IEA Counts $550bn Energy Support Bill," *Financial Times*, June 6, 2010.

51 Federative Republic of Brazil, Amazon Fund, "Purposes and Management," http://www.amazonfund.gov.br/FundoAmazonia/fam/site_en/Esquerdo /Fundo/.

52 Heinrich Böll Stiftung and the UK Overseas Development Institute, Climate Funds Update, http://www.climatefundsupdate.org/.

53 U.S.-China Joint Statement, November 17, 2009, http://www.whitehouse.gov /the-press-office/us-china-joint-statement.

54 Dan Lashoff, "Copenhagen Accord: Breakdown or Breakthrough?" NRDC *Switchboard* blog, December 19, 2009.

55 "Remarks by the President during Press Availability in Copenhagen," White House, December 18, 2009.

56 Department of Climate Change, National Development & Reform Commission of China, letter to UNFCCC executive secretary, January 28, 2010.

57 Office of the Special Envoy for Climate Change, letter to UNFCCC executive secretary, January 28, 2010.

58 European Commission, letter to UNFCCC executive secretary, January 28, 2010.

59 UNFCC, "UN Climate Change Conference in Cancún Delivers Balanced Package of Decisions, Restores Faith in Multilateral Process," December 11, 2010, http://unfccc.int/files/press/news_room/press_releases_and_advisories/application/pdf/pr_20101211_cop16_closing.pdf.

60 Hillary Rodham Clinton, press statement, December 11, 2010.

5. Follow the Money: Environmental Finance and Green Business

1 U.S. Global Change Research Program, Executive Summary, "Global Climate Change Impacts in the United States," p. 12, June 2009.

2 United States Climate Action Partnership, "A Call for Action," p. 7, January 2007.

3 BlueGreen Alliance, http://www.bluegreenalliance.org/.

4 Democratic website at Energy and Commerce of the U.S. House of Representatives, "Organizations Expressing Support for House Passage of the American Clean Energy and Security Act," December 9, 2009, http://democrats.energycommerce.house.gov/Press_111/20090624/hr2454_support passagelist.pdf.

5 H.R. 2454, American Clean Energy and Security Act of 2009, p. 1.

6 "Economics A–Z," *Economist*, http://www.economist.com/economics-a-to-z.

7 *Stern Review on the Economics of Climate Change*, launch presentation, October 30, 2006, http://webarchive.nationalarchives.gov.uk/+/http:/www.hm-treasury.gov.uk/d/stern_speakingnotes.pdf. (My emphasis in quote.)

8 Foreign Policy Association blog on Climate Change, "Cap-and-Trade vs. Carbon Tax," October 10, 2008.

9 SourceWatch, "Burson-Marsteller and Global Warming," http://www.sourcewatch.org/index.php?title=Burson-Marsteller_and_global_warming.

10 John D. Dingell, "The Power in the Carbon Tax," *Washington Post* op-ed, August 2, 2007.

11 Ibid.

12 Robert Stavins, Belfer Center for Science and International Affairs, "An Economic View of the Environment," "The Wonderful Politics of Cap-and-Trade: A Closer Look at Waxman-Markey," May 27, 2009.

13 Ibid.

14 "The Cap and Tax Fiction," *Wall Street Journal*, June 26, 2009.

15 Richard Conniff, "The Political History of Cap and Trade," *Smithsonian*, August 2009.

16 John Warner, interview for *Frontline*'s "Heat°," December 6, 2007.

17 Joe Romm, "The Failed Presidency of Barack Obama, Part 1," Climate Progress, July 22, 2010.

18 Paul Krugman, "Who Cooked the Planet?" *New York Times*, July 25, 2010.

19 Ross Douthat, "The Right and the Climate," *New York Times*, July 25, 2010.

20 Michael Tomasky, "The Senate's Energy Failure," *New York Review of Books*, July 27, 2010.

21 Robert A. Dahl, *How Democratic Is the American Constitution?* (New Haven, CT: Yale University Press, 2003), 49.

22 Ryan Lizza, "As the World Burns," *New Yorker*, October 11, 2010.

23 From *A Connecticut Yankee in King Arthur's Court*.

24 European Commission, Emissions Trading System (EU ETS) webpage, http://ec.europa.eu/clima/policies/ets/index_en.htm.

25 Point Carbon, press release, January 11, 2011.

26 Regional Greenhouse Gas Initiative, press release, September 9, 2011.

27 Pew Center on Global Climate Change, Midwest Greenhouse Gas Reduction Accord (MGGRA), http://www.pewclimate.org/what_s_being_done/in_the_states/mggra.

28 California Air Resources Board, "California Air Resources Board Gives Green Light to California's Emissions Trading Program," December 16, 2010.

29 California Air Resources Board, "California Air Resources Board Adopts Key Element of State Climate Plan," October 20, 2011.

30 Melanie Turner, "Report: Clean-Tech Powers California Economy," Silicon Valley / *San Jose Business Journal*, October 7, 2010.

31 Clean Edge, press release, December 7, 2010.

32 Felipe Calderón, video message, November 19, 2010.

33 New Zealand government, press release, December 23, 2010.

34 Australian government, press release, October 12, 2011.

35 Australian government, press release, October 24, 2011.

36 Matt Chambers, "Uncertainty over Carbon Pricing Puts Investment on Hold," *Australian*, January 10, 2011.

37 Australian government, Renewable Energy Target webpage, http://www .climatechange.gov.au/government/initiatives/renewable-target.aspx.

38 Australian government, "A New Car Plan for a Greener Future," http:// www.innovation.gov.au/INDUSTRY/AUTOMOTIVE/NEWCARPLAN /Pages/default.aspx.

39 Stuart Biggs, Chisaki Watanabe, and Mathew Carr, "Firms in Japan, Korea Oppose Start of Carbon Trading," *Bloomberg Businessweek*, December 13, 2011.

40 Peak Oil, "Japan Rejects Cap and Trade," January 1, 2011.

41 "S. Korea Sets Required Bio-diesel Mix Rate from 2012," Reuters, December 29, 2010.

42 World Bank, "State and Trends of the Carbon Market 2010," May 2010, p. 28.

43 Chris Buckley, "China Says It Is World's Top Greenhouse Gas Emitter," Reuters, November 23, 2010.

44 Barbara Finamore, "China's Carbon Intensity Target," Natural Resources Defense Council, November 27, 2009.

45 David Stanway, "China Regions to Have Binding CO2 Targets: Official," Reuters, January 13, 2011.

46 Kim Chipman and Mathew Carr, "China's Cap and Trade to Come within Five Years, Professor Stern Predicts," *Bloomberg News*, December 6, 2010.

47 David Stanway, "China Orders 7 Pilot Cities and Provinces to Set CO2 Caps," Reuters, January 16, 2012.

48 World Resources Institute, Climate Analysis Indicators Tool (CAIT), http://cait.wri.org/.

49 World Bank, "State and Trends of the Carbon Market 2010," May 2010, p. 32.

50 Ibid., pp. 34–35.

51 Trevor Curwin, "Carbon Trading May Dwarf That of Crude Oil," CNBC .com, September 29, 2009.

52 George Soros, "Seeing REDD on Climate Change," Project Syndicate, December 12, 2010.

53 Mindy Lubber, interview with author, January 13, 2011.

54 World Bank, "State and Trends of the Carbon Market 2010," May 2010, p. 2.

55 UNFCCC, "CDM Statistics," http://cdm.unfccc.int/Statistics/index.html.

56 Point Carbon, "CDM Will Grow beyond 2012 with 3.6bn CERs Issued by 2020," November 23, 2010.

57 World Bank, "State and Trends of the Carbon Market 2010," May 2010, p. 51.

58 UNFCCC, "JI Projects," http://ji.unfccc.int/JI_Projects/ProjectInfo.html.

59 Ecosystem Marketplace and Bloomberg New Energy Finance, "Building Bridges: State of the Voluntary Carbon Markets 2010," June 14, 2010., p. iv.

60 Intergovernmental Panel on Climate Change, "The IPCC Impacts Assessment," October 1990, p. 1.

61 Ibid., p. 3.

62 *Stern Review on the Economics of Climate Change*, Executive Summary, p. ii.

63 BBVA Foundation Frontiers of Knowledge Award for Climate Change, Nicholas Stern, http://www.fbbva.es/TLFU/tlfu/ing/microsites/premios /fronteras/galardonados/2010/cambioclimatico.jsp.

64 John Llewellyn, "The Business of Climate Change: Challenges and Opportunities," Lehman Brothers, February 2007, p. 1.

65 Ibid., p. 4.

66 Lord Adair Turner, chairman, UK Financial Services Authority, at Carbon Disclosure Project, "About Us," https://www.cdproject.net/en-US/Pages /About-Us.aspx.

67 World Economic Forum, press release, January 26, 2007.

68 Lubber interview.

69 Institutional Investors Group on Climate Change et al., press release, November 16, 2010.

70 Investor Network on Climate Risk, press release, January 6, 2010.

71 Al Gore, *Our Choice: A Plan to Solve the Climate Crisis* (Emmaus, PA: Rodale Press, 2009), 330.

72 Munich Re, press release, January 3, 2011.

73 Lord Peter Levene, "Catastrophe Trends and Climate Change: A Global Insurer's Perspective," http://www.lloyds.com/Lloyds/Press-Centre/Speeches /2007/01/Catastrophe_trends_and_climate_change_A_global_insurers _perspective, January 12, 2007.

74 ClimateWise, http://www.climatewise.org.uk/.

75 Zurich Financial, "The Climate Risk Challenge: The Role of Insurance in Pricing Climate-Related Risks," 2009, p. 2, http://www.zurich.com/NR

/rdonlyres/E2B5B53E-11DB-47AF-91E4-01ED6A2BDCA3/0/ClimateRisk
Challenge.pdf.

76 Evan Mills, "Ceres, from Risk to Opportunity — Insurer Responses to Climate Change," April 2009, p. 12, http://www.ceres.org/resources/reports
/insurer-responses-to-climate-change-2009.

77 Bank of America, Citi, Credit Suisse, JPMorganChase, Morgan Stanley, and
Wells Fargo.

78 The Carbon Principles, http://www.carbonprinciples.com/index.php.

79 Lubber interview.

80 Ibid.

81 Intergovernmental Panel on Climate Change, *Fourth Assessment Report*,
"Annex II — Glossary," http://www.ipcc.ch/pdf/assessment-report/ar4/syr
/ar4_syr_appendix.pdf.

82 BlueGreen Alliance, "Building the Clean Energy Assembly Line: How Renewable Energy Can Revitalize U.S. Manufacturing and the American Middle Class," November 4, 2009.

83 Nicholas Stern, "Climate: What You Need to Know," *New York Review of
Books*, June 24, 2010.

84 Nelson D. Schwartz, "Europeans Revitalize Plants to Save Jobs," *New York
Times*, February 3, 2010.

85 Daniel Schäfer, "Siemens Chief Pushes for 'Green Wave of Industrialization,'" *Financial Times*, November 9, 2009.

86 "Remarks by the President on the Economy in Schenectady, New York," January 21, 2011, White House press release.

87 Bloomberg New Energy Finance, press release, January 11, 2011.

88 Sylvia Pfeifer, "Global Deal on Climate Change Will Be the Key," *Financial
Times*, January 17, 2011.

89 Mubadala Development Company, Masdar, http://www.masdar.ae/en/home
/index.aspx.

90 Joseph Stiglitz and Nicholas Stern, "Obama's Chance to Lead the Green
Recovery," *Financial Times*, March 2, 2009.

91 Jonathan Watts, "South Korea Lights the Way on Carbon Emissions with Its
£23bn Green Deal," *Guardian*, April 21, 2009.

92 OECD, "OECD Work on Green Growth," http://www.oecd.org/document
/10/0,3746,en_2649_37465_44076170_1_1_1_37465,00.html.

93 United Nations Environment Program, press release, September 24, 2009.

94 Joel Makower and the editors of GreenBiz.com, "State of Green Business
2011," p. 3.

95 Ibid., p. 4.

96 McDonough Braungart Design Chemistry (MBDC), http://www.mbdc.com
/default.aspx.

97 P&G, "Sustainability," http://www.pg.com/en_US/sustainability/overview
.shtml.

98 Unilever, "Unilever Sustainable Living Plan," http://www.unilever.com
/sustainability/UnileverSustainableLivingPlan/index.aspx.

99 PepsiCo, "Environmental Sustainability," http://www.pepsico.com/Purpose
/Environmental-Sustainability.html.

100 Sustainability Consortium, http://www.sustainabilityconsortium.org/.

101 Tesco Corporate Responsibility Report 2011, http://www.tescoplc.com
/media/60113/tesco_cr_report_2011_final.pdf.

102 Walmart, "Sustainability," http://walmartstores.com/Sustainability/.

103 Elizabeth Sturcken, "Why Walmart's Carbon Commitment Can Make Such
a Difference," *EDF + Business*, February 25, 2010.

104 KPMG, press release, February 18, 2010.

105 Yvo De Boer, interview with author, February 4, 2011.

106 KPMG, *Corporate Sustainability: A Progress Report*, January 2011, http://
www.kpmg.com/global/en/issuesandinsights/articlespublications/pages
/corporate-sustainability.aspx.

107 McKinsey & Co., *Unlocking Energy Efficiency in the U.S. Economy*, July 2009,
http://www.mckinsey.com/Client_Service/Electric_Power_and_Natural
_Gas/Latest_thinking/Unlocking_energy_efficiency_in_the_US_economy.

108 Gilbert E. Metcalf, NBER Working Paper No. 12272, "Energy Conservation
in the United States: Understanding Its Role in Climate Policy," June 2006,
cited in "Assessing the Electric Productivity Gap and the U.S. Efficiency Opportunity," Rocky Mountain Institute, January 2009.

109 Rocky Mountain Institute, "Assessing the Electric Productivity Gap and the
U.S. Efficiency Opportunity," p. 6, January 2009.

110 Dan Reicher, interview with author, July 9, 2010.

111 Mark Ellis et al., "Do Energy Efficient Appliances Cost More?" European
Council for an Energy Efficient Economy, Summer Study 2007.

112 3M, "3P — Pollution Prevention Pays," http://solutions.3m.com/wps
/portal/3M/en_US/3M-Sustainability/Global/Environment/3P/.

113 William K. Reilly, "The Road from Rio: The Success of the Earth Summit Depends on How Well We Follow Through on Its Principles and Programs," *EPA Journal*, September/October 1992.

114 William K. Reilly, interview with author, December 10, 2010.

115 Peter Löscher, speech at Annual Shareholders' Meeting of Siemens AG, January 25, 2011.

116 Reilly interview.

6. Planet Green: The Policy, Politics, and Practice Revolution

1 Su Wei, director-general of the Department of Climate Change, in letter to UNFCCC executive secretary Yvo de Boer, January 28, 2010.

2 Barack Obama, remarks, December 18, 2009, http://www.whitehouse.gov /the-press-office/remarks-president-during-press-availability-copenhagen.

3 U.S. Department of Energy, Energy Information Administration, International Energy Statistics. http://www.eia.gov/cfapps/ipdbproject/IEDIndex3 .cfm.

4 U.S. Census Bureau, International Data Base. http://www.census.gov /population/international/data/idb/informationGateway.php/.

5 WHO, "Fuel for Life: Household Energy and Health," http://www.who.int /indoorair/publications/fuelforlife/en/index.html.

6 S. Menon et al., "Black Carbon Aerosols and the Third Polar Ice Cap," *Atmospheric Chemistry and Physics* 10 (2010): 4559–71.

7 Orville Schell, "The Message from the Glaciers," *New York Review of Books*, May 27, 2010.

8 World Bank and the State Environmental Protection Administration, People's Republic of China, "Cost of Pollution in China: Economic Estimates of Physical Damages," February 2007, p. xvii.

9 Ibid., p. xviii.

10 Ibid., p. 3.

11 Zhou Shengxian, quoted in "China Issues Warning on Climate and Growth," *New York Times*, February 28, 2011.

12 Chris Buckley, "China Vows to Cut Energy, Carbon Intensity by 2015," Reuters, February 28, 2011.

13 U.S. Department of Energy, Energy Information Administration, International Energy Statistics, http://www.eia.gov/cfapps/ipdbproject/IEDIndex3 .cfm.

14 Nan Zhou et al., "China's Energy and Carbon Emissions Outlook to 2050" (abstract), Lawrence Berkeley National Laboratory, April 2011.

15 U.S. Department of Energy, Energy Information Administration, China Brief, http://www.eia.gov/countries/cab.cfm?fips=CH.

16 International Rivers, www.internationalrivers.org, China page.

17 Ibid., "Greenhouse Gas Emissions from Dams FAQ."

18 Global Wind Energy Council, "Global Wind Capacity Increases by 22% in 2010 — Asia Leads Growth," February 2, 2011, http://www.gwec.net/index .php?id=30&no_cache=1&tx_ttnews[tt_news]=279&tx_ttnews[backPid] =4&cHash=ada99bb3b6.

19 Global Wind Energy Council, Chinese Renewable Energy Industries Association, and Greenpeace China, "China Wind Power Outlook 2010," October 1, 2010, http://www.gwec.net/fileadmin/documents/test2/wind%20 report0919.pdf.

20 Global Wind Energy Council, China page, http://www.gwec.net/index.php ?id=125.

21 Worldwatch Report no. 182: *Renewable Energy and Energy Efficiency in China: Current Status and Prospects for 2020*, October 10, 2010.

22 eSolar, press release, January 8, 2010.

23 Joseph Romm, "The Technology That Will Save Humanity," *Salon*, April 14, 2008.

24 Greenpeace International, SolarPACES, and ESTELA, "Concentrating Solar Power - Outlook 2009," May 2009, http://www.greenpeace.org/inter national/Global/international/planet-2/report/2009/5/concentrating-solar -power-2009.pdf.

25 Nathanael Baker, "Iceland and China Establish Strategic Geothermal Partnership," Energy Collective, December 23, 2010.

26 Worldwatch Institute, "China's Growth in Clean Energy Matches Ambition," October 27, 2010.

27 Tildy Bayar, "World Wind Market," *Renewable Energy World*, August 4, 2011.

28 Worldwatch Report 182, *Renewable Energy and Energy Efficiency in China: Current Status and Prospects for 2020*, October 2010, summary, http://www .worldwatch.org/bookstore/publication/worldwatch-report-182-renewable -energy-and-energy-efficiency-china-current-sta#summary.

29 Office of the U.S. Trade Representative, press release, December 22, 2010.

30 SolarWorld, press release, October 19, 2011.

31 Worldwatch Report 182, summary, http://www.worldwatch.org/bookstore /publication/worldwatch-report-182-renewable-energy-and-energy-efficiency -china-current-sta#summary.

32 U.S. Department of Energy, Energy Information Administration, Interna-

tional Energy Statistics, http://www.eia.gov/cfapps/ipdbproject/IEDIndex3
.cfm.

33 Carbon Credit Capital LLC, "India — Renewable Energy Potential," pp.
2–3, January 2011, http://www.carboncreditcapital.com/resources/press
_release/India_Article_for_Marketracker_full_v2%5B1%5D.pdf.

34 Ibid., p. 9.

35 U.S. National Renewable Energy Laboratory (NREL), REN21, et al.,
*Indian Renewable Energy Status Report: Background Report for DIREC
2010*, pp. vii–viii, October 2010, http://www.nrel.gov/docs/fy11osti/48948
.pdf.

36 Daniel Kammen, "Is the Renewable Energy Target for India within Reach?"
World Bank, February 14, 2011.

37 Gevorg Sargsyan et al., "Unleashing the Potential of Renewable Energy in
India," World Bank, p. 5, 2010.

38 Indian Ministry of New and Renewable Energy, "Resolution — JNNSM,"
January 11, 2010.

39 Desertec Foundation, Desertec–India, www.desertec-india.org.in.

40 World Bank/Energy Sector Management Assistance Program (ESMAP),
"Unleashing the Potential of Renewable Energy in India," 2011, p. 12, http://
www.esmap.org/esmap/node/1255.

41 CIA, *The World Factbook*, www.cia.gov/library, pages for India and China.

42 Global Wind Energy Council, India, http://www.gwec.net/index.
php?id=124.

43 "Unleashing the Potential of Renewable Energy in India," p. 5.

44 Hillary Clinton, Remarks at Taj Palace Hotel, Mumbai, July 18, 2009, http://
www.state.gov/secretary/rm/2009a/july/126199.htm.

45 White House, "Fact Sheet on U.S.–India Partnership on Clean Energy, Energy Security, and Climate Change," November 8, 2010.

46 "The Greenest Bail-out?" *Financial Times*, March 2, 2009.

47 South Korea Ministry of Strategy and Finance, "The Green New Deal,"
February 17, 2009.

48 "KEPCO to Invest $7.2 Billion in Smart Grid by 2030," Reuters, February 17,
2011.

49 Stacy Feldman, "Green Growth, South Korea's National Policy, Gaining
Global Attention," SolveClimate, January 26, 2011.

50 Andrew Lee, "Taiwan Initiates Energy Transition," RenewableEnergy
World.com, February 23, 2011.

51 International Energy Statistics, US Energy Information Administration, http://www.eia.gov/cfapps/ipdbproject/IEDIndex3.cfm?tid=90&pid=44 &aid=8

52 Heinrich Böll Stiftung and the UK Overseas Development Institute, Climate Funds Update, http://www.climatefundsupdate.org/.

53 Takashige Saito, "Quick Look: Renewable Energy Development in Japan," RenewableEnergyWorld.com, December 22, 2010.

54 Ernst & Young, "Renewable Energy Country Attractiveness Indices," p. 12, November 2010, http://www.ey.com/Publication/vwLUAssets/Renewable _energy_country_attractiveness_indices_-_Issue_27/$FILE/EY_RECAI _issue_27.pdf.

55 Toyota, press release, October 7, 2010.

56 Fiona Harvey, "EU Commission Report Urges 25% Carbon Emissions Cut by 2020," *Guardian*, March 8, 2011.

57 U.S. Department of Energy, Energy Information Administration, International Energy Statistics, http://www.eia.gov/cfapps/ipdbproject/iedindex3 .cfm?tid=91&pid=46&aid=31&cid=regions&syid=2000&eyid=2009&unit =MTCDPUSD.

58 Pike Research, "Energy-Efficient Buildings: Europe," March 8, 2011.

59 Pike Research, "Smart Grid Investment in Europe to Total $80 Billion by 2020," March 7, 2011.

60 John Farrell, "Distributed Generation Makes Big Numbers," Renewable EnergyWorld.com, February 28, 2011.

61 American Wind Energy Association, "U.S. Wind Industry Fast Facts," http://www.awea.org/learnabout/industry_stats/index.cfm.

62 *Frontline*, "Hot Politics," interview with William Reilly, March 19, 2007.

63 Kansas Department of Health and Environment, press release, November 18, 2007.

64 SolveClimate, "Kansas GOP Measure Could Lead to EPA Takeover of Coal Plant Permits," May 26, 2010.

65 J. Stevens, opinion in *Massachusetts et al. v. Environmental Protection Agency et al.*, 549 U.S. 497 (2007).

66 EPA, "Endangerment and Cause or Contribute Findings for Greenhouse Gases under Section 202(a) of the Clean Air Act," http://www.epa.gov /climatechange/endangerment/downloads/Federal_Register-EPA-HQ -OAR-2009-0171-Dec.15-09.pdf.

67 EPA, "EPA and NHTSA Finalize Historic National Program to Reduce

Greenhouse Gases and Improve Fuel Economy for Cars and Trucks," April 2010, http://epa.gov/otaq/climate/regulations/420f10014.pdf.

68 EPA, "Regulatory Announcement—Paving the Way toward Cleaner, More Efficient Trucks," August 2011, http://www.epa.gov/otaq/climate /documents/420f11032.pdf.

69 EPA, "Fact Sheet—Mandatory Reporting of Greenhouse Gases" (40 CFR part 98).

70 EPA, press release, December 23, 2010.

71 John Larsen, "Response to EEI's Timeline of Environmental Regulations for the Utility Industry," World Resources Institute, December 3, 2010.

72 RenewableEnergyWorld.com, "Coal-Fired Power Was the Big Loser in the Economic Downturn," June 16, 2010.

73 Progress Energy, press release, December 1, 2009.

74 "Twilight of the Coal Era?" *Green—a Blog about Energy and the Environment*, *New York Times*, June 14, 2010.

75 "Unconventional Oil and Gas 'Hot Areas,'" *Financial Times*, March 8, 2010.

76 "For Climate Relief, U.S. Will Turn to Gas," *Nature News*, June 26, 2010.

77 "How Much CO_2 Do Our Nation's Coal and Gas Plants Actually Produce?" Grist, July 6, 2009.

78 Simon Lomax, "'Massive' Closures of U.S. Coal Plants Loom, Chu Says," Bloomberg, February 9, 2011.

79 Oliver Bernstein, Sierra Club spokesman, conversation with author, March 22, 2011.

80 Mary Anne Hitt, interview with author, March 10, 2011.

81 Illinois Wind Energy Association, "About Wind Power in Illinois," http:// www.windforillinois.org/facts/.

82 Bloomberg Philanthropies, press release, July 21, 2011.

83 Kevin Parker, global head of asset management and a member of the executive committee at Deutsche Bank, quoted in an article by Steven Mufson, "Coal's Burnout: Have Investors Moved On to Cleaner Energy Sources?" *Washington Post*, January 1, 2011.

84 Ted Nace, "Ready to Rumble: A Global Movement Is Bringing Down King Coal—One Power Plant at a Time," *Earth Island Journal*, summer 2010.

85 "'Work Parties' Issue a Political Challenge: Get to Work on Climate Solutions," www.350.org, October 11, 2010.

86 Becky Tarbotton, Phil Radford, and Bill McKibben, letter in *Huffington Post*, March 7, 2011.

87 Keystone XL project page, U.S. Department of State, http://www.keystone pipeline-xl.state.gov/.

88 Edmund L. Andrews, "Bush Angers Europe by Eroding Pact on Warming," *New York Times*, March 31, 2001.

89 Christine Todd Whitman, interview for "Hot Politics," *Frontline*, January 9, 2007.

90 Philip Shabecoff, *A Fierce Green Fire: The American Environmental Movement* (Washington, DC: Island Press, 2003), 185.

91 "Canada to Phase Out Older Coal-Fired Power Plants," Planet Ark, June 28, 2010.

92 Tara Patel, "France to Shut Half Its Coal-Fed Power Plants, Curb Energy Use," Bloomberg, June 3, 2009.

93 Prime minister of Australia's website, press release, February 24, 2011.

94 Stuart McDill, "UK's Drax Coal Plant Eyes Biomass for Greener Future," Reuters, March 7, 2011.

95 "India's Coal Tax Would Generate $650 Million Annually for the Clean Energy Fund," CleanTechnica, June 27, 2010.

96 *Stern Review on the Economics of Climate Change*, http://webarchive.national archives.gov.uk/+/http:/www.hm-treasury.gov.uk/independent_reviews /stern_review_economics_climate_change/sternreview_index.cfm.

97 Barbara Finamore, "China's Announcements on Energy and Climate in Advance of Presidential Summit," NRDC *Switchboard* blog, January 18, 2011.

98 U.S. Department of Energy, Energy Information Administration, International Energy Statistics, http://www.eia.gov/cfapps/ipdbproject/IEDIndex3 .cfm?tid=90&pid=44&aid=8.

99 Barbara Finamore, "China's Twelfth Five Year Plan," NRDC *Switchboard*, March 4, 2011.

100 Kurt Kleiner, "Nuclear Energy: Assessing the Emissions," *Nature Reports Climate Change*, September 24, 2008.

101 Amory Lovins, "Four Nuclear Myths," Rocky Mountain Institute, August 13, 2009.

102 William F. Hewitt, "Nuclear Power: Running on Fumes?" reposted at Climate Progress, December 29, 2010.

103 "Tepco to Ask for $12 Billion to Help with Compensation: Sources," Reuters, November 27, 2011.

104 "Japan Nuclear Reactors Need 30-Year Fix," Associated Press, November 31, 2011.

105 "Japan: Radiation Cleanup Will Cost at Least $13 Billion, Premier Says," *Reuters*, October 21, 2011.

106 Justin McCurry, "Fukushima Released 'Twice as Much' Radioactive Material as First Thought," *Guardian*, October 28, 2011.

107 Ranjit Devraj, "Prospects Dim for India's Nuclear Power Expansion as Grassroots Uprising Spreads," *InsideClimate News*, October 25, 2011.

108 Tommy Linstroth and Ryan Bell, *Local Action: The New Paradigm in Climate Change Policy* (Lebanon, NH: University of Vermont Press / University Press of New England, 2007), x–xi.

109 Western Climate Initiative, http://www.westernclimateinitiative.org/index.php.

110 Dr. Michael Oppenheimer, interview with Charlie Rose, February 2, 2007.

111 Bill McKibben, "Warning on Warming," *New York Review of Books*, March 15, 2007.

112 Michael Grunwald, "Heroes of the Environment 2008," *Time*, September 24, 2008.

113 California Energy Commission, Energy Almanac, "California Solar Photovoltaic Statistics & Data," http://www.energyalmanac.ca.gov/renewables/solar/pv.html.

114 Go Solar California, "About the California Solar Initiative (CSI)," http://www.gosolarcalifornia.org/about/csi.php.

115 California Public Utilities Commission, Renewables/Efficiency/Environment, http://www.cpuc.ca.gov/PUC/energy/environmatters.htm.

116 City and County of San Francisco, SF Environment, http://sfgov.org/site/frame.asp?u=http://www.sfenvironment.org.

117 Natural Resources Defense Council, "Smarter Cities Project," Oakland, http://smartercities.nrdc.org/city-stories/city-profiles/large/oakland-california#tk-city-profile.

118 Andrew Nuscaren, "Los Angeles Meets 20% Renewable Energy Goal," SmartPlanet, January 18, 2011.

119 ClimateLA, Executive Summary, 2008, http://www.ci.la.ca.us/ead/pdf/ClimateLA_v5.pdf.

120 New York State Department of Environmental Conservation, "Energy/Climate Solutions," http://www.dec.ny.gov/energy/43384.html.

121 American Public Transportation Association, *2010 Public Transportation Fact Book*, April 2010, http://www.apta.com/resources/statistics/Documents/FactBook/APTA_2010_Fact_Book.pdf.

122 Metropolitan Transportation Authority, "Public Transportation for the New York Region," http://www.mta.info/mta/network.htm.

123 "Exploring Electric Vehicle Adoption in New York City," Mayor's Office of Long-Term Planning and Sustainability, January 2010.

124 Elizabeth Kolbert, "Don't Drive, He Said," *New Yorker*, May 7, 2007.

125 Transportation Alternatives, "Congestion Pricing / International Examples," http://www.transalt.org/campaigns/congestion/international.

126 Andrew Nusca, "Asia's Greenest City: Singapore," SmartPlanet, February 14, 2011.

127 Barbara Kux, quoted in Siemens, "Megacity Singapore Is Asia's Greenest City," February 14, 2011.

128 Jan Friederich, quoted in ibid.

129 "Making Cities Work — Sustainable Urban Infrastructure," Siemens presentation, February 13, 2011.

130 "Latin American Green City Index," Siemens and the Economist Intelligence Unit, November 2010, http://www.siemens.com/press/pool/de/events/corporate/2010-11-lam/Study-Latin-American-Green-City-Index.pdf.

131 Maria Gallucci, "Is Mexico City's 'Plan Verde' a Model for Latin America?" SolveClimate, November 12, 2010.

132 Institute for Transportation and Development Policy (ITDP), "Our Work — Public Transport," http://www.itdp.org/our-work/our-programs/public-transport/.

133 EMBARQ, "The Global Bus Rapid Transit (BRT) Industry," World Resources Institute, http://www.embarq.org/en/the-global-bus-rapid-transit-brt-industry.

134 UNFCC, "Project 0672 : BRT Bogotá, Colombia: TransMilenio Phase II to IV," http://cdm.unfccc.int/Projects/DB/DNV-CUK1159192623.07.

135 "European Green City Index," Siemens and the Economist Intelligence Unit, December 2009, http://www.siemens.com/press/pool/de/events/corporate/2009-12-Cop15/European_Green_City_Index.pdf.

136 ICLEI / Local Governments for Sustainability, http://www.iclei.org/.

137 The Climate Group, http://www.theclimategroup.org/.

138 Institute for Local Self-Reliance, http://www.ilsr.org/.

139 International Bank for Reconstruction and Development / The World Bank, *Cities and Climate Change: An Urgent Agenda*, Washington, DC, December 2010.

140 C40 Cities Climate Leadership Group (C40), http://live.c40cities.org/.

141 Clinton Climate Initiative, Building Retrofit Program, http://www.clinton
foundation.org/what-we-do/clinton-climate-initiative/cities/building-retrofit.

142 City of Copenhagen and ICLEI — Local Governments for Sustainability,
"The City Climate Catalogue," http://www.climate-catalogue.org/.

143 U.S. Conference of Mayors, Climate Protection Center, http://www.usmayors
.org/climateprotection/revised/.

144 Corey Dade, "In Their Own Words: GOP Candidates and Science," NPR,
September 7, 2011.

7. A Lighter Footprint: Climate Change and Sustainable Development

1 Food and Agriculture Organization, *State of the World's Forests — 2011*,
chap. 1: "The State of Forest Resources — a Regional Analysis," pp. 2–3,
http://www.fao.org/docrep/013/i2000e/i2000e01.pdf.

2 "Climate Change 2007: Mitigation; Contribution of Working Group III to
the Fourth Assessment Report of the Intergovernmental Panel on Climate
Change," chap. 9, "Forestry," p. 543, http://www.ipcc.ch/pdf/assessment
-report/ar4/wg3/ar4-wg3-chapter9.pdf.

3 Ibid., p. 544.

4 PEACE. 2007, "Indonesia and Climate Change: Current Status and Policies," 2007, pp. 1–2, http://siteresources.worldbank.org/INTINDONESIA
/Resources/Environment/ClimateChange_Full_EN.pdf.

5 Norman Myers, "Conservation of Biodiversity: How Are We Doing?" *Environmentalist*, vol. 23 (2003).

6 Carl Zimmer, "Multitude of Species Face Climate Threat," *New York Times*,
April 4, 2011; Anthony Barnosky, *Heatstroke: Nature in an Age of Global
Warming* (Washington, DC: Island Press, 2009).

7 Scott Wallace, "Last of the Amazon," *National Geographic*, January 2007.

8 Thomas E. Lovejoy, "The Amazon Forest: Let It Rain," *International Herald
Tribune*, July 27, 2006.

9 Lauren Morello and ClimateWire, "Another Amazon Drought Spurs Greenhouse Gas Emissions," *Scientific American*, February 4, 2011.

10 According to the National Snow and Ice Data Center, "In the climate system
a 'feedback loop' refers to a pattern of interacting processes where a change
in one variable, through interaction with other variables in the system, either
reinforces the original process (positive feedback) or suppresses the process
(negative feedback)."

11 British Columbia Ministry of Forests, Lands and Natural Resource Operations, "Mountain Pine Beetle," http://www.for.gov.bc.ca/hfp/mountain_pine_beetle/.

12 Barbara J. Bentz et al., "Climate Change and Bark Beetles of the Western United States and Canada: Direct and Indirect Effects," *BioScience*, September 2010, p. 603.

13 Michael G. Ryan et al., "A Synthesis of the Science on Forests and Carbon for U.S. Forests," *Issues in Ecology*, Spring 2010.

14 M. Carlson, J. Wells, and D. Roberts, "The Carbon the World Forgot," Boreal Songbird Initiative and Canadian Boreal Initiative, November 2009, pp. 3–4.

15 Peter Lee and Ryan Cheng, "Bitumen and Biocarbon: Land Use Conversions and Loss of Biological Carbon Due to Bitumen Operations in the Boreal Forests of Alberta, Canada," Global Forest Watch Canada, September 27, 2009, p. 32.

16 Corporate Ethics International, Earthworks, Natural Resources Defense Council, and Sierra Club, "Tar Sands Invasion," May 2010, http://www.nrdc.org/energy/files/TarSandsInvasion-full.pdf.

17 Rob Perks, "Appalachian Heartbreak: Time to End Mountaintop Removal Coal Mining," Natural Resources Defense Council, November 2009, p. 2.

18 Daniel C. Donato et al., "Mangroves among the Most Carbon-Rich Forests in the Tropics," *Nature Geoscience*, April 3, 2011.

19 United Nations Conference on Environment and Development (UNCED) Report, Annex III, August 14, 1992, http://www.un.org/documents/ga/conf151/aconf15126-3annex3.htm.

20 William K. Reilly, "The Road from Rio: The Success of the Earth Summit Depends on How Well We Follow Through on Its Principles and Programs," *EPA Journal*, September/October 1992.

21 William K. Reilly, interview with author, December 10, 2010.

22 Ibid.

23 The United Nations Collaborative Program on Reducing Emissions from Deforestation and Forest Degradation in Developing Countries (UN-REDD), http://www.un-redd.org/.

24 United Nations Development Program — Multi-Partner Trust Fund Office Gateway, UN-REDD, http://mptf.undp.org/factsheet/fund/CCF00.

25 International Union for Conservation of Nature (IUCN), "REDD-Plus Ex-

plained," January 6, 2011, http://www.iucn.org/about/work/programmes /forest/fp_our_work/fp_our_work_thematic/redd/redd_plus_explained/.

26 UN-REDD, http://www.un-redd.org/.

27 REDD+ Partnership, http://reddpluspartnership.org/en/.

28 United Nations Environment Program, Risoe Centre, CDM/JI Pipeline, http://cdmpipeline.org/index.htm.

29 World Bank, "The BioCarbon Fund Experience," http://siteresources.world bank.org/INTCARBONFINANCE/Resources/BioCarbon_Fund _Experience_Insights_from_AR_CDM_Projects.pdf.

30 World Bank, press release, "Africa's First Large-Scale Forestry Project under the Kyoto Protocol," March 3, 2010.

31 World Bank and UNDP, Climate Finance Options, "MDB Forest Investment Program (FIP)," http://www.climatefinanceoptions.org/cfo /node/49.

32 Forest Carbon Partnership Facility (FCPF), http://www.forestcarbon partnership.org/fcp/.

33 *2006 IPCC Guidelines for National Greenhouse Gas Inventories*, vol. 4, *Agriculture, Forestry and Other Land Use*, http://www.ipcc-nggip.iges.or.jp /public/2006gl/vol4.html.

34 Stephan Schwartzman, Daniel Nepstad, and Paulo Moutinho, "Getting REDD Right: Reducing Emissions from Deforestation and Forest Degradation (REDD) in the United Nations Framework Convention on Climate Change (UNFCCC)," Environmental Defense / Woods Hole Research Center / Instituto de Pesquisa Ambiental da Amazônia (IPAM), December 2007.

35 UNFCCC, REDD Methodologies and Tools, http://unfccc.int/methods _science/redd/methodologies/items/4538.php.

36 Global Terrestrial Observing System, http://www.fao.org/gtos/index.html.

37 Michael Lemonick, "The Great Tree Survey," *National Geographic*, May 2011.

38 U.S. Department of Agriculture, Foreign Agricultural Service, "Palm Oil: World Supply and Distribution," www.fas.usda.gov/oilseeds.

39 World Growth, "The Economic Benefit of Palm Oil to Indonesia," February 2011, http://www.worldgrowth.org/assets/files/WG_Indonesian_Palm_Oil _Benefits_Report-2_11.pdf.

40 Ecosystem Marketplace, "Norway Pledges $1 Billion for REDD as Indonesia Re-Affirms Commitment to Scheme," May 27, 2010.

41 Carbon Positive, "Indonesia Delays Deforestation Ban," January 24, 2011, http://carbonpositive2.techflaretesting.com/viewarticle.aspx?articleID =2248.

42 Olivia Rondonuwu, "Indonesia Finally Signs Forest Clearing Moratorium," Reuters, May 19, 2011.

43 Center for International Forestry Research blog, September 27, 2011, http:// blog.cifor.org/4260/norweigian-minister-for-environment-praises-indonesias -fight-against-climate-change/#.T4HOGtWibSg.

44 Roundtable on Sustainable Palm Oil (RSPO), http://www.rspo.org/.

45 Rainforest Alliance, "The Rainforest Alliance and Palm Oil," http://www .rainforest-alliance.org/agriculture/crops/palm-oil.

46 Europa, "Commission Sets Up System for Certifying Sustainable Biofuels," June 10, 2010, http://europa.eu/rapid/pressReleasesAction.do?reference =MEMO/10/247&format=HTML&aged=0.

47 "Malaysia Says Palm Oil Exports to EU Down," Agence France-Presse, March 8, 2011.

48 United Biscuits, Sustainability Report, April 2011, http://dev2.previewbox.co .uk/data/file/Sustainability%20booklet%281%29.pdf.

49 Unilever, "Sustainable Palm Oil," http://www.unilever.com/sustainability /environment/agriculture/palmoil/index.aspx.

50 Paul Sonne, "To Wash Hands of Palm Oil Unilever Embraces Algae," *Wall Street Journal*, September 7, 2010.

51 Glenn Hurowitz, "How Two 15-Year-Old Girl Scouts Are Changing a Giant Food Conglomerate," *Huffington Post*, March 23, 2011.

52 Sunanda Creagh, "Indonesia Could Cut CO_2 by 70 Pct by 2030: Report," Reuters, September 6, 2010.

53 U.S. Agency for International Development, "Forests for the Future: Sustainable Finance for Forests," fact sheet, http://www.state.gov/documents /organization/161742.pdf.

54 Rhett A. Butler, "Deforestation in the Amazon," Mongabay.com.

55 Don Hofstrand, Iowa State AgMRC Renewable Energy Newsletter, November/December 2008.

56 Union of Concerned Scientists, "Deforestation Today: It's Just Business," December 9, 2010.

57 Doug Boucher, "Brazil's Success in Reducing Deforestation," Union of Concerned Scientists, February 8, 2011.

58 Ibid.

59 Joshua Goodman, "Brazil Creates $21 Billion Fund to Slow Amazon Defor-estation," Bloomberg, August 1, 2008.

60 Amazon Fund website, "Projects," http://www.amazonfund.gov.br/Fundo Amazonia/fam/site_en/Esquerdo/Projetos/.

61 "Real-Time Evaluation of Norway's International Climate and Forest Initia-tive," Norwegian Agency for Development Cooperation, March 11, p. xiv.

62 Boucher, "Brazil's Success in Reducing Deforestation."

63 IPCC, Climate Change 2007: Working Group III: Mitigation of Climate Change — Section 9.4.3.1, "Regional Bottom-up Assessments," http://www .ipcc.ch/publications_and_data/ar4/wg3/en/ch9s9-4-3-1.html.

64 Major Economies Forum on Energy and Climate, "Declaration of the Lead-ers of the Major Economies Forum on Energy and Climate," July 2009, http://www.majoreconomiesforum.org/past-meetings/the-first-leaders -meeting.html.

65 "Joint Science Academies' Statement: Climate Change Adaptation and the Transition to a Low Carbon Society," June 2008, http://www.insaindia.org /pdf/Climate_05.08_W.pdf.

66 UN-REDD Program, "What Are the Multiple Benefits of REDD+?" http:// www.un-redd.org/AboutUNREDDProgramme/GlobalActivities/New _Multiple_Benefits/tabid/1016/Default.aspx.

67 Forest Stewardship Council, "Global FSC Certificates: Type and Distribu-tion," April 2011, http://www.fsc.org/fileadmin/web-data/public/document _center/powerpoints_graphs/facts_figures/2011-03-15-Global-FSC -Certificates-EN.pdf.

68 Rainforest Alliance, "About Us," http://www.rainforest-alliance.org/about.

69 Tensie Whelan, interview with author, April 26, 2011.

70 Rainforest Alliance, "Maya Biosphere Reserve, Guatemala," http://www .rainforest-alliance.org/adopt/projects/guatemala.

71 Whelan interview.

72 Forest Stewardship Council, "FSC Certification," http://www.fsc.org/ certification.html.

73 Jeff Hayward, interview with author, April 27, 2011.

74 UN Food and Agriculture Organization, "State of the World's Forests — 2011," p. 147, table 6: Forestry sector's contribution to employment and gross domestic product, 2006, http://www.fao.org/docrep/013/i2000e/i2000e05.pdf.

75 Ibid., chap. 1, "The State of Forest Resources — a Regional Analysis," http:// www.fao.org/docrep/013/i2000e/i2000e01.pdf.

76 Roberta Kwok, "Secondary Forests Are Worth Saving," *Nature*, January 20, 2009.

77 S. Joseph Wright and Helene C. Muller-Landau, "The Future of Tropical Forest Species," *Biotropica* 38(3), 2006.

78 The Green Belt Movement, http://greenbeltmovement.org/.

79 Ole Danbolt Mjøs, award ceremony speech, December 10, 2004.

80 United Nations Environment Program, the Billion Tree Campaign, http://www.plant-for-the-planet-billiontreecampaign.org/.

81 Dominic Casciani, "China's Floods: Is Deforestation to Blame?" BBC, August 6, 1999.

82 Jonathan Watts, "China's Loggers Down Chainsaws in Attempt to Regrow Forests," *Guardian*, March 11, 2009.

83 Ibid.

84 Al Gore, *Our Choice: A Plan to Solve the Climate Crisis* (Emmaus, PA: Rodale Press, 2009), 194.

85 U.S. Department of Agriculture, Foreign Agricultural Service, "Palm Oil: World Supply and Distribution," www.fas.usda.gov/oilseeds.

86 Watts, "China's Loggers Down Chainsaws."

87 Wangari Maathai, in *Dirt! The Movie*, 2009, dir. Gene Rosow and Bill Benenson.

88 UN Food and Agriculture Organization, *Spotlight*, "Livestock Impacts on the Environment," November 2006, http://www.fao.org/ag/magazine /0612sp1.htm.

89 Sara J. Scherr and Sajal Sthapit, "Farming and Land Use to Cool the Planet," chap. 3 in *State of the World: Into a Warming World* (WorldWatch Institute, 2006), 32.

90 Nigel Holmes, "Food for Thought," *National Geographic*, May 2011.

91 UN Food and Agriculture Organization, "Livestock Impacts on the Environment."

92 Quoted in Juliette Jowit, "UN Says Eat Less Meat to Curb Global Warming," *Observer*, September 7, 2008.

93 Interview with Rajendra Pachauri, *Progressive*, May 2009.

94 Elke Stehfest et al., "Climate Benefits of Changing Diet," Climatic Change, February 4, 2009.

95 Agronomy is a branch of agriculture dealing with field-crop production and soil management.

96 IPCC, Climate Change 2007: Working Group III: Mitigation of Climate

Change — Section 8.4.1.1, "Cropland Management," http://www.ipcc.ch
/publications_and_data/ar4/wg3/en/ch8s8-4-1-1.html.

97 U.S. Department of Agriculture, Farm Service Agency, "Conservation Reserve Program Celebrates 25 Years," December 23, 2010.

98 Scherr and Sthapit, "Farming and Land Use to Cool the Planet," 37.

99 Anna Lappé, interview with author, April 15, 2011.

100 Organic Farming Research Foundation, http://ofrf.org/.

101 U.S. Department of Agriculture, National Agricultural Statistics Service, "New USDA Data Offers In-Depth Look at Organic Farming," February 3, 2010, http://www.nass.usda.gov/Newsroom/2010/02_03_2010.asp.

102 UN Food and Agriculture Organization, "Introducing Organic Research Centre Alliance," November 2009, http://www.fao.org/fileadmin/templates/organicag/files/ORCA_en.pdf.

103 J. N. Pretty et al., "Resource-Conserving Agriculture Increases Yields in Developing Countries," *Environmental Science & Technology*, December 21, 2005.

104 Bob Watson et al., International Assessment of Agricultural Science and Technology for Development (IAASTD) presentation, London, April 15, 2008.

105 IAASTD, Beverly D. McIntyre et al., eds., *IAASTD Global Report: Agriculture at a Crossroads*, April 2008, p. 401, http://www.agassessment.org/reports/IAASTD/EN/Agriculture%20at%20a%20Crossroads_Global%20Report%20%28English%29.pdf.

106 Office of the United Nations High Commissioner for Human Rights (OHCHR), "'Agroecology Outperforms Large-Scale Industrial Farming for Global Food Security,' Says UN Food Expert," June 22, 2010, http://www.srfood.org/images/stories/pdf/press_releases/20100622_press_release_agroecology_en.pdf.

107 Anna Lappé, *Diet for a Hot Planet: The Climate Crisis at the End of Your Fork and What You Can Do about It* (New York: Bloomsbury, 2010), 149.

108 "Plan B: Mobilizing to Save Civilization," episode 12 of PBS's *Journey to Planet Earth*.

109 U.S. Department of State, press release, September 21, 2010.

110 Amanda Leigh Haag, "The Even Darker Side of Brown Clouds," *Nature Reports Climate Change*, September 2007.

111 Solar Cookers International, http://www.solarcookers.org.

112 Nevin Cohen, interview with author, April 28, 2011.

113 Growing Power, http://www.growingpower.org/.

114 Tom Philpott, "Norman Borlaug, Meet Will Allen," Grist, September 28, 2009.

115 La Via Campesina, http://viacampesina.org/en/.

116 Lappé interview.

117 U.S. Department of Agriculture, "Know Your Farmer, Know Your Food," http://www.usda.gov/wps/portal/usda/usdahome?navid=KNOWYOUR FARMER.

118 Kathleen Merrigan, "Food Hubs: Creating Opportunities for Producers across the Nation," USDA blog, April 19, 2011.

119 RUAF Foundation (Resource Centre on Urban Agriculture and Food Security), http://www.ruaf.org/.

120 Lappé interview.

121 Let's Move, http://www.letsmove.gov/.

122 Rainforest Alliance, "Our Work in Sustainable Agriculture," http://www.rainforest-alliance.org/work/agriculture.

123 Sustainable Agriculture Network, Mission and Goals, http://sanstandards.org/sitio/subsections/display/1.

124 Jeff Hayward, interview with author, April 27, 2011.

125 Tensie Whelan, "Vision for Sustainability," Rainforest Alliance, http://www.rainforest-alliance.org/sites/default/files/publication/pdf/vision_brochure.pdf.

8. A Resilient Future: Adaptation, Education, Law, and Lifestyle

1 *Stern Review on the Economics of Climate Change*, Executive Summary, October 2006, p. 5, http://webarchive.nationalarchives.gov.uk/20100407171416/http:/www.hm-treasury.gov.uk/d/Executive_Summary.pdf.

2 Report brief, "Ecological Impacts of Climate Change," Committee on Ecological Impacts of Climate Change of the National Research Council, 2008.

3 Climate Change 2007: Impacts, Adaptation and Vulnerability; Contribution of Working Group II to the Fourth Assessment Report of the Intergovernmental Panel on Climate Change, Summary for Policymakers, pp. 8–9, http://www.ipcc.ch/pdf/assessment-report/ar4/wg2/ar4-wg2-spm.pdf.

4 UN Environment Program / World Glacier Monitoring Service, "Global Glacier Changes: Facts and Figures," Summary, September 2008, http://www.grid.unep.ch/glaciers/pdfs/summary.pdf.

5 National Snow and Ice Data Center, "All about Glaciers," http://nsidc.org /cryosphere/glaciers/index.html.

6 Orville Schell, "The Message from the Glaciers," *New York Review of Books*, May 27, 2010.

7 Suzanne Goldenberg, "World's Glaciers Melting at Accelerated Pace, Leading Scientists Say," *Guardian*, January 20, 2010.

8 Climate Change 2007: Impacts, Adaptation and Vulnerability; Contribution of Working Group II to the Fourth Assessment Report of the Intergovernmental Panel on Climate Change, chap. 10, "Asia," p. 484, http://www.ipcc .ch/pdf/assessment-report/ar4/wg2/ar4-wg2-chapter10.pdf.

9 Mauri S. Pelto, "Glaciers and Hydropower," Nichols College, http://www .nichols.edu/departments/glacier/glacier%20runoff%20hydropower.htm.

10 "An Overview of Glaciers, Glacier Retreat, and Subsequent Impacts in Nepal, India and China," WWF Nepal Program, March 2005, p. 3.

11 "Global Glacier Changes: Facts and Figures," p. 27.

12 Climate Change 2007: Working Group I: The Physical Science Basis, "FAQ 5.1, Is Sea Level Rising?" http://www.ipcc.ch/publications_and_data/ar4 /wg1/en/faq-5-1.html.

13 Susan Solomon et al., "Irreversible Climate Change Due to Carbon Dioxide Emissions," *PNAS* (*Proceedings of the National Academy of Sciences*) 106, no. 6 (2009): 1704–9; published online January 28, 2009.

14 Climate Change 2007: Impacts, Adaptation and Vulnerability; Contribution of Working Group II to the Fourth Assessment Report of the Intergovernmental Panel on Climate Change, chap. 6, "Coastal Systems and Low-Lying Areas," p. 338, http://www.ipcc.ch/pdf/assessment-report/ar4/wg2/ar4-wg2 -chapter6.pdf.

15 Ibid., table 6.7: Health effects of climate change and sea-level rise in coastal areas, p. 335.

16 Ibid., table 6.8: Key hotspots of societal vulnerability in coastal zones, p. 337.

17 Solomon et al., "Irreversible Climate Change."

18 Heidi Cullen, *The Weather of the Future: Heat Waves, Extreme Storms, and Other Scenes from a Climate-Changed Planet* (New York: Harper, 2010), 3.

19 Insurance Information Institute, "Climate Change: Insurance Issues," January 2011, http://www.iii.org/issues_updates/climate-change-insurance-issues .html.

20 Evan Lehmann, "Inland Storms, Growing in Violence, Drive Insurers to Accept Riskier Reality," *New York Times*, May 20, 2011.

21 ClimateWise, the Geneva Association, MCII, and UNEP Finance Initiative, "Global Insurance Industry Statement on Adapting to Climate Change in Developing Countries," September 6, 2010, http://www.climate-insurance .org/upload/pdf/201009_Global_Insurance_Industry_Group_Statement_on _Adaptation_in_Developing_Countries.pdf.

22 Dan Haugen, "Risky Business," *Momentum*, Spring 2011.

23 UNFCCC, "Fact Sheet: The Need for Adaptation," http://unfccc.int/press /fact_sheets/items/4985.php.

24 UNFCCC, Nairobi Work Programme, Partners and Action Pledges, http://unfccc.int/adaptation/nairobi_work_programme/partners_and _action_pledges/items/5005.php.

25 UNFCCC, "Fact Sheet: The Need for Adaptation."

26 Global Environment Facility, GEF-Administered Trust Funds, http://www .thegef.org/gef/trust_funds.

27 Adaptation Fund, "Adaptation Fund Board Approves Project for Financing and Accredits Two Implementing Entities," September 16, 2011, http://www .adaptation-fund.org/media/adaptation-fund-board-approves-project -financing-and-accredits-two-implementing-entities.

28 Heinrich Böll Stiftung and the UK Overseas Development Institute, Climate Funds Update, Types of Funding, http://www.climatefundsupdate.org /themes.

29 *Report of the Secretary-General's High-Level Advisory Group on Climate Change Financing*, November 5, 2010, p. 5, http://www.un.org/wcm/webdav /site/climatechange/shared/Documents/AGF_reports/AGF%20Report.pdf.

30 Athena Ballesteros et al., "Summary of Developed Country 'Fast-Start' Climate Finance Pledges," World Resources Institute, May 20, 2011.

31 Athena Ballesteros, "The Road to the Green Climate Fund," World Resources Institute, February 14, 2011.

32 Climate Change 2007: Impacts, Adaptation and Vulnerability; Contribution of Working Group II to the Fourth Assessment Report of the Intergovernmental Panel on Climate Change, chap. 17, "Assessment of Adaptation Practices, Options, Constraints and Capacity," p. 719, http://www.ipcc.ch/pdf /assessment-report/ar4/wg2/ar4-wg2-chapter17.pdf.

33 "Facing the Consequences," *Economist*, November 25, 2010.

34 U.K. Environment Agency, "The Thames Barrier," http://www .environment-agency.gov.uk/homeandleisure/floods/38353.aspx.

35 Netherlands, Delta Program, http://www.deltacommissaris.nl/english/.

36 Cullen, *Weather of the Future*, 125.

37 Ibid., 127.

38 Bryson Bates et al., "Climate Change and Water — Executive Summary," IPCC, June 2008, p. 4.

39 Executive Summary, New York City Panel on Climate Change, *Climate Change Adaptation in New York City: Building a Risk Management Response*, May 18, 2010.

40 Cullen, *Weather of the Future*, 256.

41 Leslie Kaufman, "A City Prepares for a Warm Long-Term Forecast," *New York Times*, May 22, 2011.

42 ICLEI, Climate Adaptation, http://www.iclei.org/index.php?id=10832.

43 Urban Climate Change Research Network, http://uccrn.org/.

44 UN-Habitat, "Cities and Climate Change: Global Report on Human Settlements 2011," http://www.unhabitat.org/pmss/listItemDetails.aspx?publicationID=3086.

45 ICLEI — Second World Congress on Cities and Adaptation to Climate Change, http://resilient-cities.iclei.org/bonn2011/program/.

46 World Bank and C40 Cities, "Unique Cooperation Will Address Key Barriers Cities Face in Obtaining Financing for Climate-Related Projects," June 1, 2011.

47 Centers for Disease Control, Climate Change and Public Health, http://www.cdc.gov/climatechange/default.htm.

48 U.S. Global Change Research Program, "What We Do," http://www.globalchange.gov/what-we-do.

49 Paul R. Epstein, "Health and Climate Change: 7 Ways You Are Being Harmed," *Atlantic*, September 26, 2011.

50 White House, Climate Change Adaptation Task Force, http://www.whitehouse.gov/administration/eop/ceq/initiatives/adaptation.

51 U.S. Agency for International Development, "Climate Change Pillar: Adaptation," http://www.usaid.gov/our_work/environment/climate/policies_prog/adaptation.html.

52 UNEP, Climate Change Adaptation, http://www.unep.org/climatechange/adaptation/.

53 UNEP, "High Mountain Glaciers and Climate Change: Challenges to Human Livelihoods and Adaptation," December 7, 2010, pp. 24–29, http://www.unep.org/pdf/himalayareport_screen.pdf.

54 International Centre for Integrated Mountain Development (ICIMOD),

"Glacial Lake Mapping and Glacial Lake Outburst Flood (GLOF) Risk Assessment in the Hindu Kush–Himalayas," www.icimod.org/?q=235.

55 Red Cross / Red Crescent Climate Centre, http://www.climatecentre.org/site /home.

56 Katie Fehrenbacher, "The Hot New Sector in Greentech: Adaptation," *GigaOm*, December 12, 2010.

57 Barbara Kiser, "Moveable Feast: The Floating Gardens of Bangladesh," International Institute for Environment and Development, April 2010.

58 "Facing the Consequences," *Economist*, November 25, 2010.

59 The Climate Reality Project, http://presenters.climaterealityproject.org/.

60 U.S. Green Building Council, http://www.usgbc.org/Default.aspx.

61 World Green Building Council, http://www.worldgbc.org/site2/.

62 UNFCCC, National Communications Annex I, http://unfccc.int/national _reports/annex_i_natcom_/items/1095.php.

63 Fifth U.S. Climate Action Report, "Chapter 9: Education, Training and Outreach," U.S. Department of State, May 28, 2010, http://unfccc.int/ resource/docs/natc/usa_nc5.pdf.

64 Australia's Fifth National Communication on Climate Change, February 12, 2010, p. 169, http://unfccc.int/resource/docs/natc/aus_nc5.pdf.

65 The Fifth National Communication of France to the United Nations Framework Convention on Climate Change (English abstract), November 2009, http://unfccc.int/resource/docs/natc/franc5abs.pdf.

66 Japan's Fifth National Communication under the United Nations Framework Convention on Climate Change, January 2010, p. 337, http://unfccc.int /resource/docs/natc/jpn_nc5.pdf.

67 IGLO (International Action on Global Warming), http://astc.org/iglo/.

68 American Museum of Natural History, press release, October 2008.

69 Association for the Advancement of Sustainability in Higher Education. http://www.aashe.org/.

70 U.S. Green Building Council, Center for Green Schools, http://centerfor greenschools.org/home.aspx.

71 Ibid.

72 Second Nature, "Colleges and Universities Report Significant Progress in Confronting Climate Disruption," April 5, 2011.

73 Michael Gerrard, interview with author, May 11, 2011.

74 Climate Literacy and Energy Awareness Network (CLEAN), http://cleanet .org/index.html.

75 Climate Change Litigation in the U.S., Arnold & Porter LLP, as of April 2, 2012, http://www.climatecasechart.com/.

76 Non-U.S. Climate Change Litigation Chart, Center for Climate Change Law, as of April 2, 2012, http://www.climatecasechart.com/.

77 Gerrard interview.

78 Ibid.

79 Climate Change Litigation in the U.S., Arnold & Porter LLP, October 27, 2011.

80 Opinion of the U.S. Supreme Court in *American Electric Power Co., Inc. v. Connecticut*, 564 U. S. _____ (2011), p. 14.

81 Climate Justice Campaign, Center on Race, Poverty and the Environment, http://www.crpe-ej.org/crpe/index.php/campaigns/climate-justice.

82 Gerrard interview.

83 Gallup, "In Major Economies, Many See Threat from Climate Change," July 8, 2009, http://www.gallup.com/poll/121526/Major-Economies-Threat -Climate-Change.aspx?CSTS=tagrss.

84 Gallup, "Fewer Americans, Europeans View Global Warming as a Threat," April 20, 2011, http://www.gallup.com/poll/147203/Fewer-Americans -Europeans-View-Global-Warming-Threat.aspx.

85 World Public Opinion, "Multi-Country Poll Reveals That Majority of Peo- ple Want Action on Climate Change, Even if it Entails Costs," December 3, 2009, http://worldpublicopinion.org/pipa/articles/btenvironmentra/649 .php?lb=bte&pnt=649&nid=&id=.

86 "Americans' Global Warming Beliefs and Attitudes in May 2011," Yale Proj- ect on Climate Change Communication and Center for Climate Change Communication, June 6, 2011, http://environment.yale.edu/climate/news /americans-global-warming-beliefs-and-attitudes-in-may-2011/.

87 Yale Project on Climate Change Communication, press release, February 24, 2010.

88 Jeff Tollefson, "Educators Take Aim at Climate Change," *Nature News*, Oc- tober 26, 2010.

89 National Religious Partnership for the Environment, http://www.nrpe.org /index.html.

90 Coalition on the Environment and Jewish Life, http://coejl.org/.

91 Catholic Climate Covenant, http://catholicclimatecovenant.org/.

92 Benedict XVI, World Day of Peace message, January 1, 2010.

93 National Council of Churches of Christ, Eco-Justice Program Office, http:// nccecojustice.org/climate/.

94 Evangelical Environmental Network, http://creationcare.org/index.php.

95 Greg Haegele, "'Creation Care': A Growing Movement," *Treehugger*, July 3, 2008.

96 Creation Care for Pastors, http://www.creationcareforpastors.com/.

97 The Office of His Holiness the Dalai Lama, press release, December 7, 2008.

98 Jared Diamond, "What's Your Consumption Factor?" *New York Times*, January 2, 2008.

99 Ibid.

100 Climate Change 2007: Working Group III: Mitigation of Climate Change, figure SPM.1, "Global Warming Potential (GWP) weighted global green house gas emissions 1970–2004," http://www.ipcc.ch/publications_and_data /ar4/wg3/en/figure-spm-1.html.

101 Isabelle DePommereau, "In Germany, a Different Kind of Electronic Revolution," *Christian Science Monitor*, June 3, 2011.

102 Ryan Brown, "'Losing Our Cool': The High Price of Staying Cool," *Salon*, July 5, 2010.

103 MarketsandMarkets, "Global Geothermal Power and Heat Pump Market Outlook (2010–2015)," January 2011, http://www.marketsandmarkets.com /Market-Reports/geothermal-energy-market-284.html.

104 Environmental Defense Fund, "Study Concludes Information-Based Energy Efficiency Can Save Americans Billions," May 23, 2011, http://www.edf.org /news/study-concludes-information-based-energy-efficiency-can-save -americans-billions.

105 OPOWER, http://opower.com/what-is-opower.

106 Tomoko A. Hosaka, "Japan Asks Workers to Dress Cool to Save Energy," Associated Press, June 5, 2011.

107 U.S. Postal Service, "Sustainability," http://about.usps.com/what-we-are -doing/green/welcome.htm.

108 Rainforest Alliance, "Our Work in Sustainable Tourism," http://www .rainforest-alliance.org/work/tourism.

109 The International Ecotourism Society, http://www.ecotourism.org/.

110 Global Sustainable Tourism Council, http://new.gstcouncil.org/.

111 United Nations Environment Program, Global Partnership for Sustainable Tourism, http://www.unep.fr/scp/tourism/activities/partnership /Documents/2011Factsheet_FINAL.pdf.

112 United Nations Environment Program, International Task Force on Sustain-

able Tourism Development, http://www.unep.fr/scp/tourism/activities/task force/index.htm.

113 *Back to the Future: State of the Voluntary Carbon Markets 2011*, Ecosystem Marketplace and Bloomberg New Energy Finance, June 2011, p. 8.

114 Verified Carbon Standard, http://www.v-c-s.org/how-it-works/vcs-program.

115 *Back to the Future: State of the Voluntary Carbon Markets 2011*, Ecosystem Marketplace and Bloomberg New Energy Finance, June 2, 2011, p. 11.

116 Carbonfund.org, http://www.carbonfund.org/.

117 Josie Garthwaite, "Bike-Share Schemes Shift into High Gear," *National Geographic News*, June 7, 2011.

118 *Guardian*, Datablog, September 2, 2009.

119 David Batty and David Adam, "Vegetarian Diet Is Better for the Planet, Says Lord Stern," *Guardian*, October 26, 2009.

120 "Plan B: Mobilizing to Save Civilization," PBS's *Journey to Planet Earth*, http://www.pbs.org/journeytoplanetearth/plan_b/index.html.

index

acid rain, 5, 27, 79, 90, 112–13, 116, 151, 167–68

adaptation: in Bangladesh, 244–45; Cancún Adaptation Framework, 104; by cities, 239–41; funding for, 101–2, 104, 161, 172, 235–37, 256; and insurance industry, 233–34; planning and knowledge sharing, 242–44; and public health, 241–42; to sea-level rise and storms, 180, 237–38, 239; and sustainable development, 208, 245; and UNFCCC, 82–83, 235; and water supply, 238–39

afforestation, 196, 199, 212, 214

agriculture: GHG emissions from, 24, 91, 257; influence of livestock, 214–16; mitigation potential of, 216–18; palm oil, 201–4; as principal driver of deforestation, 191, 214–16; public awareness, 225–26; sustainable agriculture, 133, 139, 218–21; urban agriculture, 222–24; and water, 229–30, 231–32

air conditioning, 54, 65, 67, 74, 145, 258–59

air pollution: in Asia, 151; from biomass burning and forest fires, 150, 201; in California, 97, 177–79; in China, 27, 151–52, 154; clean tech for reducing, 29, 46; and endangerment finding, 165–66; as environmental justice issue, 254–55; international treaty on, 79; in London, 26; from power plants, 49, 169; natural gas for reducing, 168; in New York City, 181. *See also* acid rain

algae, 29, 47, 140, 204

allowances, 111, 117. *See also* cap and trade

Amazon rainforest, 47, 190; deforestation in, xvii, 45, 62, 205–7; and drought, 191–92; funding to preserve, 101, 206

American Recovery and Reinvestment Act (ARRA), 63, 70, 95. *See also* green stimulus

Australia: adaptation conference, 244; bilateral agreements, 101; cap and trade, 92, 120, 172; Great Barrier Reef, 232; marine energy, 42; politics in, 91–92, 120; public education, 248; solar power, 37, 55; support for renewable energy, 120

auto industry, 93, 96–97, 161, 165–66

Bali Roadmap, 76

Bangladesh, 231, 244–45

Ban Ki-moon, 76, 77

Bank of America Tower, 67, 73, 142, 180, 259

bilateral agreements, 100–2, 136, 236

biochar, 46–47, 221

biofuel: impact on food prices, 45; impacts from land-use changes, 45, 201, 203, 205; for military use, 97–98; from palm oil, 201, 203; second-generation biofuels, 45–46, 144

biomass: in developing world, 27, 52, 150–51, 154, 157, 221–22; power from, 29, 44–47, 161, 172–73. *See also* algae; biochar; biofuel; black carbon

climate funds, 100–2, 104, 159, 161, 197–99, 206, 235–37, 261–62

Climategate, xvi, 1–2, 76

Climate Project, 23, 246–47

climate risk, xvii, 109, 119, 127–34, 141–42, 146, 170, 235. *See also* adaptation; insurance industry

climate science, xvi, 1–9, 11, 15–18, 78, 90, 98, 99, 248. *See also* education and public awareness

Climatic Research Unit (CRU), 1–2, 3

Clinton, Hillary, 94, 99, 104, 158–59, 221

Clinton, President William J., 84, 110, 164. *See also* Clinton Climate Initiative

Clinton Climate Initiative (CCI), 132, 186

coal: carbon dioxide emissions, 24–26, 149–50, 154; decline in use, 30, 40–41, 123, 131–32, 149, 163–64, 167–70, 172–74, 251–53; economic losses from, 27–28, 50; environmental impacts, 26–27, 195; in Europe, 91; health impacts from, 26–27, 49, 50, 150; mine safety, 27; as special interest, 16, 112, 114, 115–16, 166–67. *See also* Goodell, Jeff

co-benefits, 125, 134, 197

cogeneration, 29, 45, 53–54, 157, 168

Cohen, Nevin, 223, 224, 225

combined heat and power (CHP). *See* cogeneration

common but differentiated responsibilities. *See* Kyoto Protocol

Commoner, Barry, xviii; *Making Peace with the Planet*, 11; *The Politics of Energy*, 41, 168

concentrating solar power (CSP), 34, 35–37, 38, 66, 154–55. *See also* Desertec Initiative

congestion pricing, 181–82

consumption divide, 256–57

conversion loss, 50–51, 53

Conway, Erik M. (*Merchants of Doubt*), 7, 15

cookstoves, 221–22

Copenhagen, 10, 185. *See also* COP 15

COP 15 (Fifteenth Session of the Conference of the Parties to the United Nations Framework Convention on Climate Change), xvi, 2, 76–78, 99, 102–3, 121, 141, 148, 152, 171; funding commitments at, 236

COP 16. *See* Cancún Agreements

cryosphere, 228–31. *See also* glaciers

Cullen, Heidi (*The Weather of the Future*), 233, 243–44

De Boer, Yvo, 77, 141–42

decentralized energy (DE). *See* distributed generation

deforestation, xv, xvii, 6, 24, 121–22, 132, 190–97; in the Amazon, xvii, 45, 62, 205–7; as driven by agriculture, 191; as driven by biofuel production, 45, 201, 203, 205; funding for avoided deforestation, 101, 126, 197–200. *See also* palm oil; Reducing Emissions from Deforestation and Degradation

demand-side management (DSM), 65–69, 160, 239, 259–60

Denmark, 20, 54, 90, 125–26, 172, 249

Department of Agriculture (USDA), 94, 218, 219, 224

Department of Defense (DOD), 47, 53, 56, 69, 97–98

Department of Energy (DOE), 14, 38, 47, 55–56, 74, 94, 133, 142, 153; advanced research, 70, 97. *See also* Chu, Steven; Energy Information Administration

Department of the Interior, 32, 94, 98

Department of Transportation (DOT), 96, 165–66

Desertec Initiative, 35–37, 66, 67–68, 155, 157–58

United Nations Conference on Environment and Development (1992). *See* Earth Summit

United Nations Environment Program (UNEP), 7, 14, 79, 81, 137, 200, 243, 244, 261

United Nations Food and Agriculture Organization (FAO), 197, 200, 211, 214–15, 220

United Nations Framework Convention on Climate Change (UNFCCC), 2, 10, 14–15, 76, 109, 123, 141, 156, 208, 256; and adaptation, 235–37; and education, 247; founding of, 7–8, 9, 195–96; structure of, 82–83, 123, 125. *See also* Bali Roadmap; Cancún Agreements; Clean Development Mechanism; COP 15; Kyoto Protocol; Reducing Emissions from Deforestation and Degradation

United Nations General Assembly, 8, 77

U.S. Global Change Research Program (USGCRP), 85, 98, 106, 241–42, 248

U.S. Green Building Council (USGBC), 72–73, 74, 247, 249–50, 259

U.S. Senate, 82, 84, 92–94, 106–7, 108, 110, 111–16, 166

Vaitheeswaran, Vijay V., 62, 64, 70

Villach Conference, 7–8

voluntary carbon markets, 121, 122, 126, 261–62

Walmart, 60, 133, 134, 140–41, 168

waste-to-energy (WTE), 20

Watt, James, 24, 28

Waxman, Henry, 93, 107, 110

Waxman-Markey bill (American Clean Energy and Security Act), 93, 106–8, 110–11, 112, 113, 115

websites and blogs (covering climate, clean tech, and sustainability), 14–15, 17–18, 20–21, 138, 223, 246, 250

Western Climate Initiative (WCI), 119, 168–69, 177

Whelan, Tensie, 209–10, 226

white roofs, 71, 73–74

wind power, 31–34, 40–41, 48, 67, 154, 155, 157, 163–64; equipment manufacturing, 57, 146, 155, 158. *See also* microwind

World Bank, 13, 27, 158, 186, 220, 243; climate and energy funding, 102, 156–57, 199, 235–36, 241. *See also* Stern, Lord Nicholas

World Business Council on Sustainable Development, 71, 132, 145

World Meteorological Organization (WMO), 7, 200

Yergin, Daniel, 28, 62